Genetic and Evolutionary Computation

Series Editors:
David E. Goldberg, ThreeJoy Associates, Inc., Urbana, IL, USA
John R. Koza, Stanford University, Los Altos, CA, USA

More information about this series at http://www.springer.com/series/7373

Rick Riolo • Bill Worzel • Brian Goldman
Bill Tozier
Editors

Genetic Programming Theory and Practice XIV

Springer

Editors
Rick Riolo
Center for the Study of Complex Sys
University of Michigan
Ann Arbor, MI, USA

Brian Goldman
Colorado State University
Fort Collins, CO, USA

Bill Worzel
Evolution Enterprises
Ann Arbor, MI, USA

Bill Tozier
Ann Arbor
MI, USA

ISSN 1932-0167
Genetic and Evolutionary Computation
ISBN 978-3-030-07300-8 ISBN 978-3-319-97088-2 (eBook)
https://doi.org/10.1007/978-3-319-97088-2

This Springer imprint is published by the registered company Springer Nature Switzerland AG
The registered company address is: Gewerbestrasse 11, 6330 Cham, Switzerland

Preface

This is the written record of the 14th meeting of the annual Genetic Programming Theory and Practice Workshop, which was hosted by the Center for the Study of Complex Systems at the University of Michigan, in Ann Arbor, May 19–21, 2016.

It is, as a matter of course, woefully incomplete. It can serve only as a fragmentary record of our meeting. I say this not as an apology, but rather as a sort of gentle warning, of a sort I rarely seem to see in these collections.

Let me try to explain. The central and explicit focus of the GPTP Workshop has always been the conversations that are fostered *at* the meeting itself. These conversations happen among the invited keynote speakers, the technical participants, the students, and sponsors who spend somewhat more than 3 days in a room together. The technical work that invited speakers have prepared *before* they arrive, and which they bring with them to present in session and discuss, is really just the provocation or *seed* of the "real workshop." As a result, the proceedings volume from each workshop should only ever be read as a record of where some subset of attendees started individually, not as a position in which we ended up as a group by the end. The *real* results will only gradually appear in the subsequent literature, as a cloud of works that may not even explicitly refer to this meeting, and in the subtly changed directions of ongoing research programs in years to come.

In other words, think of these chapters as the *inputs* to a years-long dynamical process, not as the final *output* of that process.

Further, this volume does not even manage to cover all of the presentation-driven portion of our discussions. It is a sad but unavoidable truth of modern worklife that only a fraction of our most influential participants have the spare time to produce a chapter for you to read. Our three invited keynote speakers, for example, inevitably have a strong influence on our collective attention, setting the tone and focus of the entire week as soon as they begin speaking... but they are rarely able to make time to provide written contributions here. And because we so actively seek out both industrial and academic speakers, we often hear at least one or two presentations that treat contemporary professional work from an industrial domain. Proprietary and ongoing projects cannot always be translated successfully into an academic-style paper.

Since this was our 14th meeting, you should suspect that even calling these chapters the "inputs" of a long-term process misses the preceding history. If we can only detect the "outputs" in the workshop's future-facing light-cone—by seeing the consequences of the week's conversations and insights in the attendees' actions—then surely almost all of 2016s workshop was in turn influenced by earlier sessions. You will find chapters here describing the ongoing efforts of Michael Affenzeller's HEAL group, as they implement, explore, and distribute ideas and algorithms in their HeuristicLab system, which have in several cases arisen directly from earlier GPTP workshops. Michael Korns carries forward his many years' effort in support of industrial-strength symbolic regression systems. There are at least three chapters here from Lee Spector, Nic McPhee, Thomas Helmuth, and colleagues, which have arisen from many years of fruitful exploration of the Push language for GP, and so on. I doubt a single chapter lacks a strong link from some earlier GPTP conversation.

Therefore, think of this book as a collection of blurred snapshots, taken from the center of the web of conversations that make up the "real" GPTP Workshop. The historian interested in the context from which they have arisen should review the prior 13 volumes in the series *at least* as diligently as she reviews the more mundane published conference literature in adjacent years.

Assume that we took the time to speak to one another—substantively, but not *constantly*—about the material presented in this volume. But in the moment and in the room together, our conversations were much more along the lines of, "That's fascinating, but have you considered...," or "I wonder if you've noticed that what you did in Section 3 could be related to what Smith did in her work on...," and so forth. We do not just clap and grumble at one another—we discuss. When it works, this workshop is a *generative process*, not a ritualized presentation. Most of the "work" it does in the field will have happened after the attendees arrived, in the lunch breaks, the hallway conversations, and the notes we have scribbled on a napkin in the pub afterward, or in our notebooks in hotel rooms after quiet thoughtful dinners.

In the following 14 chapters, you will find the subset of contributions from invited speakers who were able to provide them. They have done a good job contextualizing and framing their work, subject to the caveats I have spelled out above. So rather than simply revisiting each one in turn, let me try to fill in the gaps between them a bit, beginning with our three keynote speakers.

Dr. Joanna Masel, from Ecology and Evolutionary Biology at the University of Arizona, was the first of our invited keynote speakers. She spoke on "Evolution of molecular error rates, and the consequences for robustness, evolvability, and the de novo emergence of new protein-coding genes from junk DNA." She pointedly reminded the audience—we are for the most part computer science folks—of the deep fundamental differences between biological and "computational" evolutionary processes. In the course of her presentation, she did an excellent job conveying both the fascinating complexity of biological evolution (and evolutionary biology, the discipline), and also the potential shortcomings of our greatly simplified mental models in evolutionary computing. Throughout the workshop, she was able to helpfully remind many speakers and participants of the ways in which such an overly glib simple model might lead one into a proverbial ditch.

Dr. Stephanie Forrest, from the University of New Mexico and the Santa Fe Institute (a colleague of several attendees), spoke in her keynote lecture about "Software: Evolution, Robustness, and Diversity (also, the Mutation Cliff)." She spoke about her group's and colleagues' ongoing research into the complexities of *real* (which is to say: human-written) software systems and in particular their work in the new field of genetic improvement and automated software repair. As with Dr. Masel's effort, we in the audience were frequently and pointedly reminded of the "simplifying assumptions" our work often makes for the sake of being tractable. Once again, there are many ways in which the consequences from our overly simplified framing notions can stumble, when faced with the externalities of the real world. In particular, she spoke of her own practical and philosophical explorations of what "fitness" might mean in the context of producing well-repaired software originally produced by human programmers. As my notes have it, "What is a *reasonable* way of quantifying the performance of a broken or repaired piece of low-level software infrastructure?" We tend in the field of GP not to think very often about bug reports, side effects, and other matters that live past the interface with software development, deployment, and usability... but our more advanced work inevitably bumps up against it.

On the third day of our meeting, Dr. Cosma Shalizi from the Statistics Department at Carnegie Mellon University and the Santa Fe Institute spoke on "Bayesian Learning, Evolutionary Search, and Information Theory." And boy did he. He pointed out remarkable (but as far as I am aware, previously unremarked) similarities in the deep structure of Bayesian learning representations and algorithms, the replicator equation and other core dynamical models from theoretical population biology, and the ways in which information (in Shannon's sense) is "handled" by these processes. In other words, in a bit more than an hour, he stitched together three increasingly independent disciplinary approaches to learning and dynamical systems models and described a strong framework for exploring what it might mean for evolution to "learn." This delightful presentation—which several of us hope he will have published somewhere soon—brought in frameworks from disciplines that will certainly benefit from an ontological reconciliation like this. In other words, we look forward to reading his paper on the matter *at least as much* as you do.

Intermingled with the three keynote speakers' talks, and the presentations from the invited speakers whose chapters follow, there were also innumerable breakfast, lunch, and dinner meetings (all ad hoc of course). Watch for their effects in the *future* work of the attendees, as you read the works they brought to the table to begin our conversations with one another.

Ann Arbor, MI, USA — William Tozier
Fort Collins, CO, USA — Brian Goldman
Ann Arbor, MI, USA — Rick Riolo
Ann Arbor, MI, USA — William P. Worzel
December 2017

Acknowledgments

We feel obliged not only to *thank* the participants in the 2016 GPTP Workshop but to emphasize how proud we are to have played our little role in bringing them together. We realize (because they say it over and over, year after year) that the guests think this is an enjoyable and inspiring meeting, but the pleasure is literally ours—at least in hindsight. If you listen, you will notice our anecdotes all talk about the way we watched you talk with one another so productively and pleasantly, not just what you said but how you *were* together. So thank you all for being interesting people who can now and then listen to one another.

The keynote speakers are not just punctuation marks or particularly honored members of our little community, but have been tasked with a crucial role of jostling our attention in a productive way and steering us away from our discipline-killing tendency for *only ever* talking to one another. Our speakers in 2016 did an admirable job in this, and we appreciate their company at least as much as the insights they brought from adjacent disciplines.

We could not undertake these ambitious workshops without the generous sponsorships, with whose help we are increasingly able to bring new colleagues into the fold. In 2016, these included:

- The Center for the Study of Complex Systems at the University of Michigan and especially Carl Simon and Charles Doering, the champions of the workshop series
- John Koza
- Jason Moore
- Babak Hodjat at Sentient
- Michael Korns and Gilda Cabral
- Mark Kotanchek at Evolved Analytics
- The Heuristic and Evolutionary Algorithms Laboratory at the Upper Austria University of Applied Science

Finally, there are those without whose help we would not have been able to manage: Linda Wood and Mita Gibson at the Center for the Study of Complex Systems.

Thank you all for making this possible, and a pleasure.

Contents

Contributors

Michael Affenzeller Heuristic and Evolutionary Algorithms Laboratory, University of Applied Sciences Upper Austria, Hagenberg, Austria

Institute for Formal Models and Verification, Johannes Kepler University, Linz, Austria

Itay Azaria Department of Computer Science, Ben-Gurion University, Beer-Sheva, Israel

Wolfgang Banzhaf Department of Computer Science, Memorial University, St. John's, NL, Canada

BEACON Center of the Study of Evolution in Action, Michigan State University, East Lansing, MI, USA

Armand R. Burks BEACON Center for the Study of Evolution in Action, Michigan State University, East Lansing, MI, USA

Bogdan Burlacu Heuristic and Evolutionary Algorithms Laboratory, University of Applied Sciences Upper Austria, Hagenberg, Austria

Institute for Formal Models and Verification, Johannes Kepler University, Linz, Austria

Maggie M. Casale University of Minnesota, Morris, Morris, MN, USA

Mauro Castelli NOVA IMS, Universidade Nova de Lisboa, Lisbon, Portugal

Achiya Elyasaf Department of Computer Science, Ben-Gurion University, Beer-Sheva, Israel

Mitchell D. Finzel University of Minnesota, Morris, Morris, MN, USA

Philipp Fleck Heuristic and Evolutionary Algorithms Laboratory, University of Applied Sciences Upper Austria, Hagenberg, Austria

Steven Gustafson Maana, Bellevue, WA, USA

Thomas Helmuth Computer Science, Washington and Lee University, Lexington, VA, USA

Erik Hemberg MIT CSAIL, Cambridge, MA, USA

Babak Hodjat Sentient Technologies, San Francisco, CA, USA

Chris Holmes University of Oxford, Oxford, UK

Ting Hu Department of Computer Science, Memorial University, St. John's, NL, Canada

Perla S. Juárez-Smith Tecnológico Nacional de México/I.T. Tijuana, Tijuana, B.C., Mexico

Michael Kommenda Heuristic and Evolutionary Algorithms Laboratory, University of Applied Sciences Upper Austria, Hagenberg, Austria

Institute for Formal Models and Verification, Johannes Kepler University, Linz, Austria

Michael F. Korns Analytic Research Foundation, Henderson, NV, USA

Krzysztof Krawiec Institute of Computing Science, Poznan University of Technology, Poznań, Poland

Gabriel Kronberger Heuristic and Evolutionary Algorithms Laboratory, University of Applied Sciences Upper Austria, Hagenberg, Austria

Pierrick Legrand Universitè de Bordeaux, Institut de Mathèmatiques de Bordeaux, UMR CNRS 5251, Bordeaux, France

CQFD Team, Inria Bordeaux Sud-Ouest, Talence, France

Nicholas Freitag McPhee Division of Science and Mathematics, University of Minnesota, Morris, Morris, MN, USA

Risto Miikkulainen Sentient Technologies, San Francisco, CA, USA

Jason H. Moore Institute for Biomedical Informatics, University of Pennsylvania, Philadelphia, PA, USA

Luis Muñoz Tecnológico Nacional de México/I.T. Tijuana, Tijuana, B.C., Mexico

Lawrence Murray University of Oxford, Oxford, UK

Luiz Otavio V. B. Oliveira DCC, Universidade Federal de Minas Gerais, Belo Horizonte, Brazil

Randal S. Olson Institute for Biomedical Informatics, University of Pennsylvania, Philadelphia, PA, USA

Una-May O'Reilly MIT CSAIL, Cambridge, MA, USA

Fernando E. B. Otero School of Computing, University of Kent, Chatham Maritime, UK

Gisele L. Pappa DCC, Universidade Federal de Minas Gerais, Belo Horizonte, Brazil

William F. Punch BEACON Center for the Study of Evolution in Action, Michigan State University, East Lansing, MI, USA

Jacob Rosen MIT CSAIL, Cambridge, MA, USA

Oliver Schütze Computer Science Department, CINVESTAV-IPN, Mexico City, Mexico

Hormoz Shahrzad Sentient Technologies, San Francisco, CA, USA

Saul Shanabrook Computer Science, University of Massachusetts, Amherst, MA, USA

Sara Silva BioISI Biosystems and Integrative Sciences Institute, Faculty of Sciences, University of Lisbon, Lisbon, Portugal

Moshe Sipper Department of Computer Science, Ben-Gurion University, Beer-Sheva, Israel

Lee Spector Cognitive Science, Hampshire College, Amherst, MA, USA

Arun Subramaniyan BHGE - Digital, San Ramon, CA, USA

Jerry Swan Department of Computer Science, University of York, York, UK

Leonardo Trujillo Tecnológico Nacional de México/I.T. Tijuana, Tijuana, B.C., Mexico

Leonardo Vanneschi NOVA IMS, Universidade Nova de Lisboa, Lisbon, Portugal

Stephan M. Winkler Heuristic and Evolutionary Algorithms Laboratory, University of Applied Sciences Upper Austria, Hagenberg, Austria

Institute for Formal Models and Verification, Johannes Kepler University, Linz, Austria

Aisha Yousuf Eaton Corporation, Southfield, MI, USA

Emigdio Z-Flores Tecnológico Nacional de México/I.T. Tijuana, Tijuana, B.C., Mexico

Chapter 1
Similarity-Based Analysis of Population Dynamics in Genetic Programming Performing Symbolic Regression

Stephan M. Winkler, Michael Affenzeller, Bogdan Burlacu, Gabriel Kronberger, Michael Kommenda, and Philipp Fleck

Abstract Population diversity plays an important role in the evolutionary dynamics of genetic programming (GP). In this paper we use structural and semantic similarity measures to investigate the evolution of diversity in three GP algorithmic flavors: standard GP, offspring selection GP (OS-GP), and age-layered population structure GP (ALPS-GP). Empirical measurements on two symbolic regression benchmark problems reveal important differences between the dynamics of the tested configurations. In standard GP, after an initial decrease, population diversity remains almost constant until the end of the run. The higher variance of the phenotypic similarity values suggests that small changes on individual genotypes have significant effects on their corresponding phenotypes. By contrast, strict offspring selection within the OS-GP algorithm causes a significantly more pronounced diversity loss at both genotypic and, in particular, phenotypic levels. The pressure for adaptive change increases phenotypic robustness in the face of genotypic perturbations, leading to less genotypic variability on the one hand, and very low phenotypic diversity on the other hand. Finally, the evolution of similarities in ALPS-GP follows a periodic pattern marked by the time interval when the bottom layer is reinitialized with new individuals. This pattern is easily noticed in the lower layers characterized by shorter migration intervals, and becomes less and less noticeable on the upper layers.

S. M. Winkler (✉) · M. Affenzeller · B. Burlacu · M. Kommenda
Heuristic and Evolutionary Algorithms Laboratory, University of Applied Sciences Upper Austria, Hagenberg, Austria

Institute for Formal Models and Verification, Johannes Kepler University, Linz, Austria
e-mail: Stephan.Winkler@fh-hagenberg.at

G. Kronberger · P. Fleck
Heuristic and Evolutionary Algorithms Laboratory, University of Applied Sciences Upper Austria, Hagenberg, Austria

R. Riolo et al. (eds.), *Genetic Programming Theory and Practice XIV*, Genetic and Evolutionary Computation,
https://doi.org/10.1007/978-3-319-97088-2_1

Keywords Genetic programming · Symbolic regression · Population dynamics · Genetic diversity · Phenotypic diversity · Offspring selection · Age-layered population structure · ALPS

1.1 Introduction: Genetic Programming, Population Diversity, and Population Dynamics

Genetic Programming (GP) [6, 9] is a powerful optimization technique which evolves a population of tree-encoded solution candidates according to the rules of natural selection. Similar to its biological counterpart, the GP algorithm is dependent on the two steps of Darwinian evolution: variation (due to crossover and mutation) and selection. Genotypes (G) are mapped into phenotypes (P), an evaluation function $f : S \rightarrow \mathbb{R}$ (where S is the solution space) assigns a fitness value to each individual in the population. If a variation in a trait is more successful (i.e., it improves an organism's propagation success rate by allowing it to have more viable offspring) then that trait may eventually come to dominate the population. When certain phenotypic traits dominate the population to the detriment of others, genotypic variation with a high adaptive potential but lower fitness runs the risk of becoming extinct as a consequence of selection. Schaper and Louis [10] suggest this happens when the more "globally fit" do not have time to be found or to fix in the population over evolutionary timescales. The authors suggest that "strong biases in the rates at which traits can arrive through variation may direct evolution towards outcomes that are not simply the fittest". Thus, loss of diversity has a negative impact on the search by reducing the population's adaption potential.

While low genotypic diversity will in most cases hinder the genetic process to generate novel solution candidates, low phenotypic diversity might indicate that there is no significant search progress since newly created individuals are not better than their parents. This is why we here specifically analyze phenotypic as well as genotypic diversities, as this shall also enable a more detailed discussion about the reasons of premature convergence (seen in genotypic diversity) and its consequences (seen in phenotypic diversity).

In this paper, we analyze empirically the loss of population diversity for different GP flavors and symbolic regression problem instances. We introduce computational methods for measuring diversity at the genotypic and phenotypic level, and investigate the correlation between the two. Burke et al. [2] provide a good overview of various distance measures, analyzing the correlation between fitness and diversity; structural versus evaluation based solutions similarity analysis for symbolic regression was for example discussed in [16]. We show the progress of genotype and phenotype population diversity for three GP algorithmic configurations, namely standard GP, GP with strict offspring selection, and ALPS-GP.

The chapter is organized as follows: Sect. 1.2 describes the tree distance metrics that were used for similarity calculation and the methodology for our experiments. Section 1.3 describes the test settings, Sect. 1.4 summarizes the obtained results, and in Sect. 1.5 we give our conclusions.

1.2 Similarity Measures

We here introduce a new genotype similarity measure based on the bottom-up tree distance [11] and a phenotype similarity measure based on the correlation between two individuals' outputs.

Since our similarity measures are symmetrical, the number of similarity calculations necessary to compute the average similarity for a population of N individuals is $\frac{N(N-1)}{2}$. Therefore, the population diversity is given by:

$$Div(T) = 1 - \frac{\sum_{i=1}^{N-1} \sum_{j=i+1}^{N} Sim(t_i, t_j)}{N(N-1)/2}, \tag{1.1}$$

where $Sim(t_1, t_2)$ can be either the bottom-up or the phenotypic similarity.

1.2.1 Genotypic Similarity

Genotypic similarity is calculated using a measure similar to the tree edit distance, called the *bottom-up distance*. The bottom-up tree distance is a flexible distance measure based on the largest common forest between trees, as described by Valiente [11]. It has the advantage of maintaining the same time complexity, namely linear in the size of the two trees regardless of whether the trees are ordered or unordered. The algorithm works as follows:

1. In the first step, it computes the compact directed acyclic graph representation G of the largest common forest $F = t_1 \uplus t_2$ (consisting of the disjoint union between the two trees). The graph G is built during a bottom-up traversal of F (in the order of non-decreasing node height). Two nodes in F are mapped to the same vertex in G if they are at the same height and their children are mapped to the same sequence of vertices in G. The bottom-up traversal ensures that children are mapped before their parents, leading to $O(|t_1|+|t_2|)$ time for adding vertices in G corresponding to all nodes in F. This step returns a map $K : F \rightarrow G$ which is used to compute the bottom-up mapping.
2. The second step iterates over the nodes of t_1 in level-order and builds a mapping $M : t_1 \rightarrow t_2$ using K to determine which nodes correspond to the same vertices in G. The level-order iteration guarantees that every largest unmapped subtree of t_1 will be mapped to an isomorphic subtree of t_2. Finally, the bottom-up distance between trees t_1 and t_2 is calculated as

$$\text{BottomUpDistance}(t_1, t_2) = \frac{2 \times |M(t_1, t_2)|}{|t_1| + |t_2|}. \tag{1.2}$$

Thus, the similarity of t_1 and t_2 is defined as

$$\text{GenotypicSimilarity}(t_1, t_2) = 1 - \text{BottomUpDistance}(t_1, t_2). \tag{1.3}$$

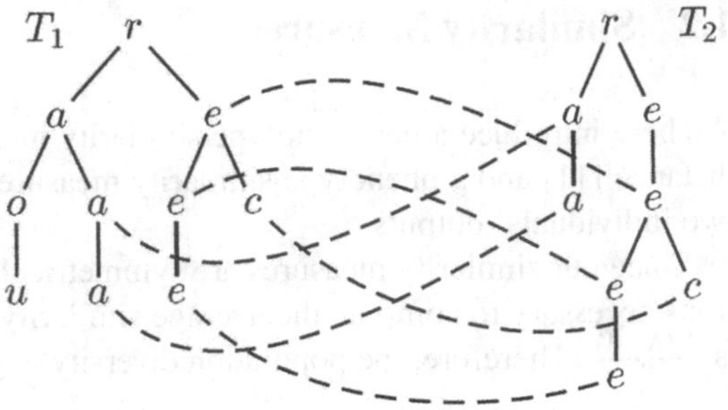

Fig. 1.1 Bottom-up mapping between two trees t_1 and t_2 (see [11])

By taking two times the size of the bottom-up mapping between the two trees, we make sure that the similarity values will always fall inside the [0, 1] interval (Fig. 1.1).

1.2.2 Phenotypic Similarity

Similarity at the phenotype level is calculated with regard to an individual's response on the training data. Individuals with the same response (with the same *semantics*) are considered phenotypically similar regardless of their actual structure.

In this paper, we introduce a phenotypic similarity measure based on the squared Pearson product-moment correlation coefficient:

$$R^2_{X,Y} = \left(\rho_{X,Y}\right)^2 = \left(\frac{\mathrm{Cov}(X,Y)}{\sigma_X \sigma_Y}\right)^2. \tag{1.4}$$

Since $\rho \in [-1, +1]$, the R^2 correlation coefficient will always return a similarity value in the interval [0, 1]. One pitfall of using the above formula is that individuals with a constant response cannot be compared, as the Pearson correlation coefficient cannot be calculated when the variance is zero. In this special case, we consider two individuals with constant response to be completely similar to each other (returning a similarity value of 1). Thus, the phenotypic similarity measure is calculated using the formula:

$$\mathrm{PhenotypicSimilarity}(t_1, t_2) = \begin{cases} 1 & \text{if } \mathrm{Var}(t_1) = \mathrm{Var}(t_2) = 0 \\ 0 & \text{if } \mathrm{Var}(t_1) = 0 \text{ or } \mathrm{Var}(t_2) = 0 \\ R^2_{t_1,t_2} & \text{otherwise} \end{cases} \tag{1.5}$$

1.3 Test Setup

For analyzing the effects of selection mechanisms and algorithmic settings on GP population dynamics we ran test series using standard GP, GP with offspring selection and ALPS-GP. As benchmark data sets we used the *Poly-10* and the *Tower* data sets. For our test series we used the implementations of these algorithms and problems in HeuristicLab [14, Ver. 3.13], an open source framework for heuristic optimization that can be retrieved from http://dev.heuristiclab.com/.

The parameters for all here used algorithms were set such that they represent typical as well as competitive settings for the problem instances, i.e. typical configurations that are frequently used in practical applications and theoretical research studies.

1.3.1 Algorithms

1.3.1.1 Standard Genetic Programming (SGP)

First we applied symbolic regression using genetic programming as implemented in HeuristicLab. The following parameter settings were chosen for these tests:

- Population size: 500 individuals
- Termination criterion: 1000 generations
- Tree initialization: probabilistic Tree Creation (PTC2) [7]
- Maximum tree size: 50 nodes, 10 levels
- Elites: 1 individual
- Parent selection: tournament selection, group size 5
- Crossover: subtree crossover, 100% probability
- Mutation: 25% mutation rate, each mutation is performed either as single-point, multi-point, remove branch or replace branch mutation
- Fitness function: coefficient of determination R^2 [3]
- Terminal symbols: constant, weight * variable
- Function symbols: binary functions ($+, -, \times, \div$, exp, log)

1.3.1.2 Genetic Programming with Offspring Selection (OSGP)

Secondly, we used GP with strict offspring selection (OS) as explained in [1]. OS-GP shifts the focus of selection towards adaptive change by introducing an additional selection step where newly created individuals are accepted into the population only if their fitness exceeds that of their parents. The algorithm produces as many individuals as needed in order to fill in a new generation of individuals. In this context, the *active selection pressure* is defined as the ratio between the total number of produced offspring and the number of individuals needed to fill a generation (i.e., the population size). The active selection pressure varies every generation depending on how easy it is to generate better offspring. The active selection pressure at generation i is expressed as:

$$SelectionPressure(i) = \frac{|GeneratedOffspring(i)|}{|SuccessfulOffspring(i)|} = \frac{|GeneratedOffspring(i)|}{|Population|}. \tag{1.6}$$

We use the selection pressure as termination criterion, i.e., the algorithm is terminated as soon as the selection pressure reaches a predefined maximum value.

Most parameters for these OS-GP tests are equal to those used for standard GP; OS-GP specific parameter settings were set as follows:

- Population size: 200 individuals
- Termination criterion: Maximum selection pressure 200
- Parent selection: Gender specific [13]; proportional and random
- Offspring selection: Strict, i.e. success ratio = 1.0 and comparison factor = 1.0 [1]

1.3.1.3 ALPS GP

Age-layered population structure (ALPS) GP uses a novel measure of age to separate the population into multiple layers [4]. Each layer states a maximum age so that lower layers contain younger individuals and higher layers contain older individuals.

An individual's age determines how long it is allowed to remain in its current layer. Randomly generated individuals start with an age of zero, while in the default age-inheritance scheme, individuals generated by crossover inherit the age of the oldest parent plus one. Other inheritance schemes, such as using the younger or the average of the parents' age, have also been studied [5].

The age concept allows younger, less fit individuals to compete for survival within their own age layer, without being dominated by already matured individuals. A fair competition allows reseeding the lowest layer with new randomly generated individuals during the run, increasing the overall genetic diversity.

We used ALPS-GP as implemented in HeuristicLab and used the same operators and settings for tree size, initialization, crossover, and mutation as in standard GP and OS-GP. The following ALPS specific parameter settings were chosen for these tests:

- Population size: 300 individuals
- Age inheritance: Older
- Replacement strategy: Comma
- Aging: Age gap 20, polynomial aging scheme, i.e. the first layer (layer 0) is newly initialized every 20 generations, and individuals may move to upper layers at generations 20, 40, 80, and 160

1.3.2 Problem Instances

We tested the aforementioned GP algorithms on two benchmark regression problems to examine population dynamics. The problems were taken from the recommended GP benchmark problems [15] and are both available within the HeuristicLab framework.

- The *Poly-10* data set [8] consists of 500 samples with ten variables $x_{1...10}$ and the response variable y. The values $x_{1...10}$ were generated by randomly (uniformly) drawing values from the interval $[-1, +1]$, the response values were calculated according to the following equation:

$$y = f(\mathbf{x}) = x_1x_2 + x_3x_4 + x_5x_6 + x_1x_7x_9 + x_3x_6x_{10}$$

- The *Tower* data set [12] comes from an industrial problem on modeling gas chromatography measurements of the composition of a distillation tower. It contains 5000 records and 25 potential input variables, the response variable is the propylene concentration at the top of the distillation tower. The samples were measured by a gas chromatograph and recorded as floating averages every 15 min. The 25 potential inputs are temperatures, flows, and pressures related to the distillation tower. The *Tower* data set can be downloaded from http://www.symbolicregression.com/?q=towerProblem.

1.4 Test Results

Similarity and quality measurements were averaged over ten runs for each problem instance (*Poly-10* and *Tower*) and algorithmic configuration.

Figures 1.2 and 1.3 show the evolution of best and average quality and similarity values for standard GP. We notice that genotypic similarity remains at a constant level on both test problems. On the other hand, phenotypic similarity and average quality are higher on the *Tower* problem, suggesting a correlation between the two.

The distribution of similarity values per generation is displayed in Fig. 1.4a as 2d histograms, measured every 100 generations on the *Poly-10* problem. In the charts, the x-axis represents phenotypic similarity while the y-axis represents genotypic similarity. The results reveal that genotype similarity increases at a higher rate than the phenotypic similarity. The presence of multiple "islands" on the phenotypic similarity axis (at the same genotype similarity level) suggests that individuals in the population are organized into different semantic groups, some consisting of highly similar individuals.

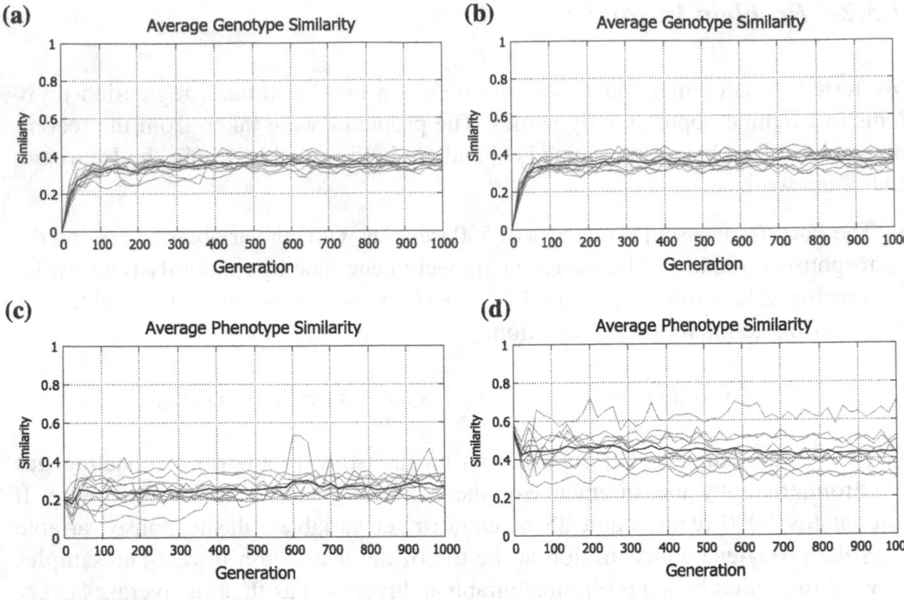

Fig. 1.2 Genotypic and phenotypic population diversity in standard GP, on the *Poly-10* and *Tower* problems. Thick lines represent average values over ten repetitions. (**a**) *Poly-10* genotypic similarity. (**b**) *Tower* genotypic similarity. (**c**) *Poly-10* phenotypic similarity. (**d**) *Tower* phenotypic similarity

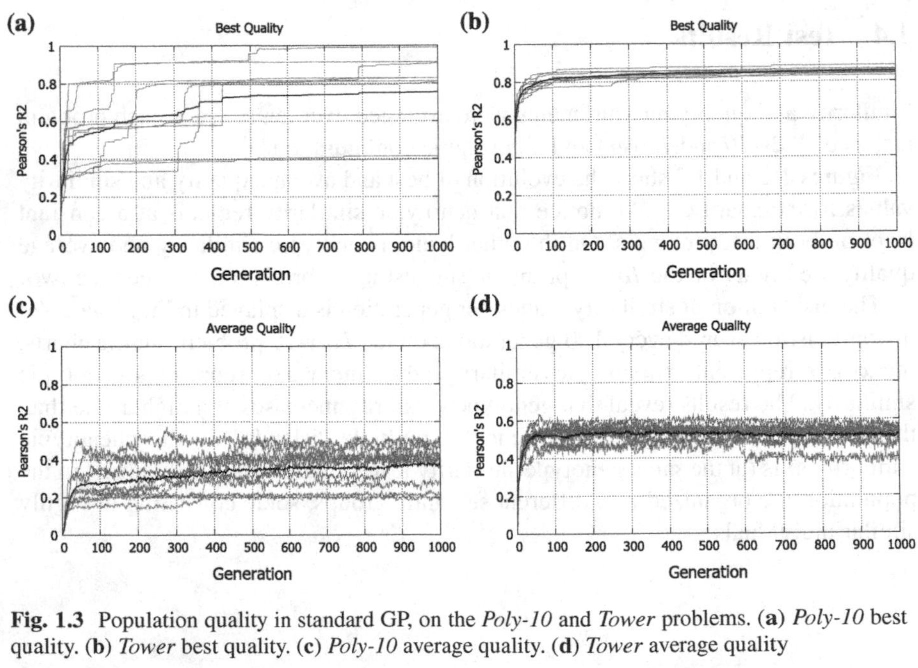

Fig. 1.3 Population quality in standard GP, on the *Poly-10* and *Tower* problems. (**a**) *Poly-10* best quality. (**b**) *Tower* best quality. (**c**) *Poly-10* average quality. (**d**) *Tower* average quality

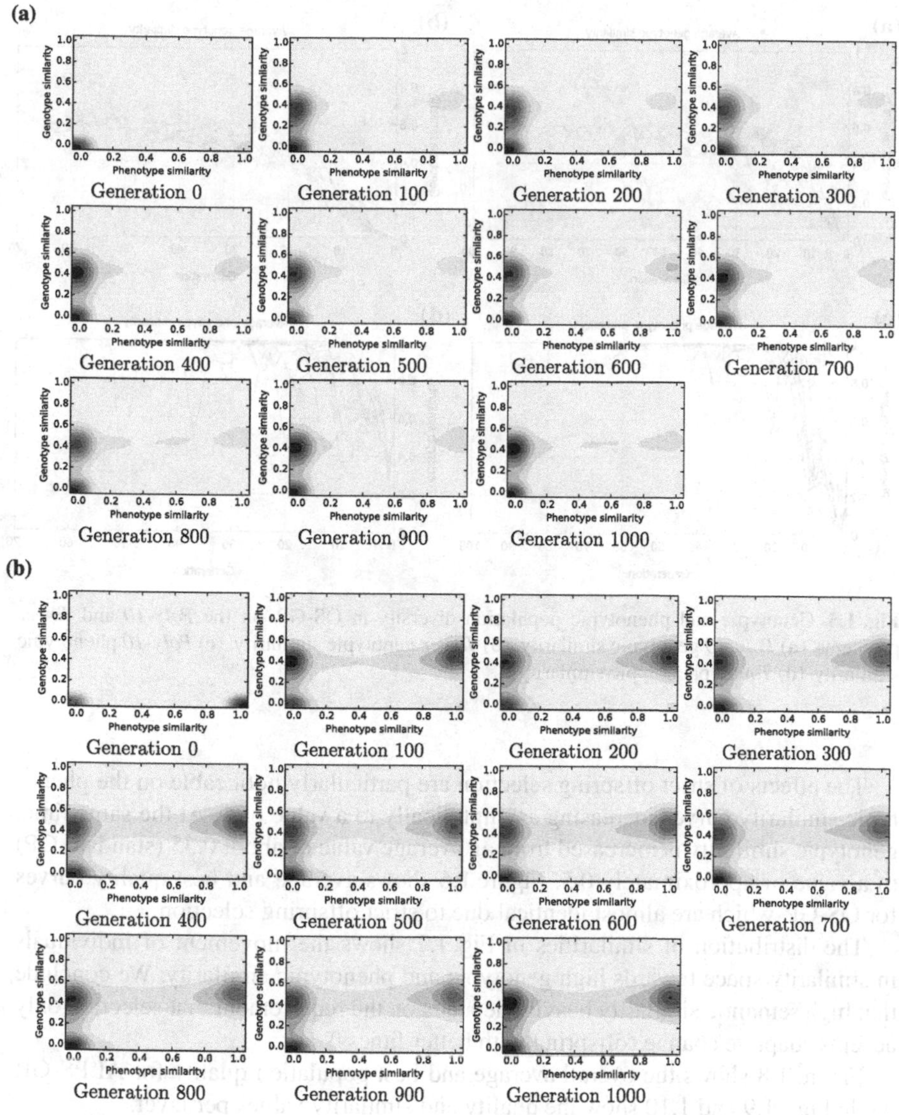

Fig. 1.4 Distribution of genotypic vs. phenotypic similarities in standard GP. (**a**) *Poly-10* problem. (**b**) *Tower* problem

We compare standard GP similarities with those measured on the OS-GP runs. Figure 1.5 indicates a steeper increase of similarity levels (both genotypic and phenotypic) towards significantly higher values.

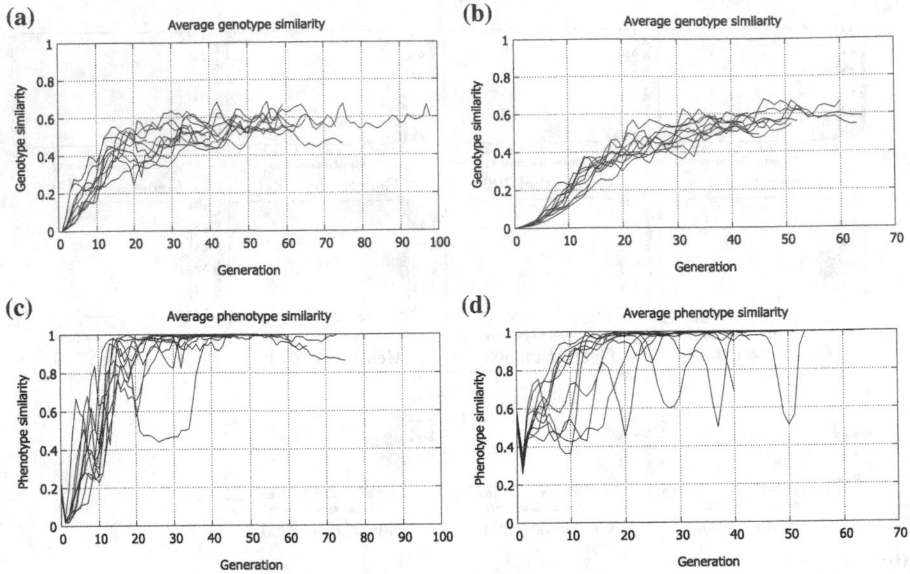

Fig. 1.5 Genotypic and phenotypic population diversity in OS-GP, on the *Poly-10* and *Tower* problems. (**a**) *Poly-10* genotypic similarity. (**b**) *Tower* genotypic similarity. (**c**) *Poly-10* phenotypic similarity. (**d**) *Tower* phenotypic similarity

The effects of strict offspring selection are particularly noticeable on the phenotypic similarity curves increasing asymptotically to a value of 1. At the same time, genotypic similarity is increased from an average value of about 0.35 (standard GP) to a value of approximately 0.5. Figure 1.6 shows average and best quality curves for OS-GP, which are almost identical due to strict offspring selection.

The distribution of similarities in Fig. 1.7 shows the movement of individuals in similarity space towards high genotypic and phenotypic similarity. We conclude that high semantic similarity heavily depends on the requirement that selection only accepts adaptive change (offspring with better fitness).

Figure 1.8 shows the overall average and best population quality for ALPS-GP, while Figs. 1.9 and 1.10 show the quality and similarity values per layer.

We notice that ALPS-GP is able to achieve a better average best quality than standard GP despite the fact that the overall average population quality is lower than the corresponding standard GP value, due to the lower-quality of the bottom ALPS layers. As expected, each age layer displays the same similarity behavior as standard GP, with the average population similarity increasing in the intervals

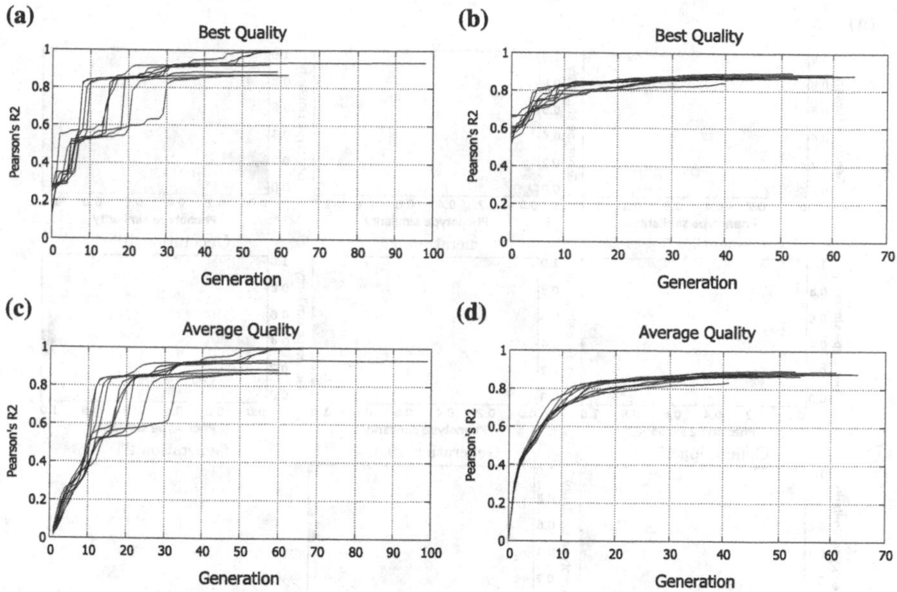

Fig. 1.6 Population quality in OS-GP, on the *Poly-10* and *Tower* problems. (**a**) *Poly-10* best quality. (**b**) *Tower* best quality. (**c**) *Poly-10* average quality. (**d**) *Tower* average quality

between reinitialization (for layer 0) or migration to the upper layers (for layer $\geq$ 1). As new genetic variation propagates slower to the older layers (due to larger migration intervals and competition in the lower layers), these are characterized by both higher average qualities and higher similarities. The relationship between average quality and average phenotypic similarity for standard GP and the upper layers of ALPS suggests a correlation between quality and phenotypic similarity.

1.5 Conclusion

Concerning the analysis of algorithms dynamics we have seen clearly that the progress of the populations' similarity is dramatically different when using different flavors of GP algorithms for symbolic regression where a conscious distinction between phenotypic and genotypic diversity has the potential to offer additional insights in terms of population dynamics.

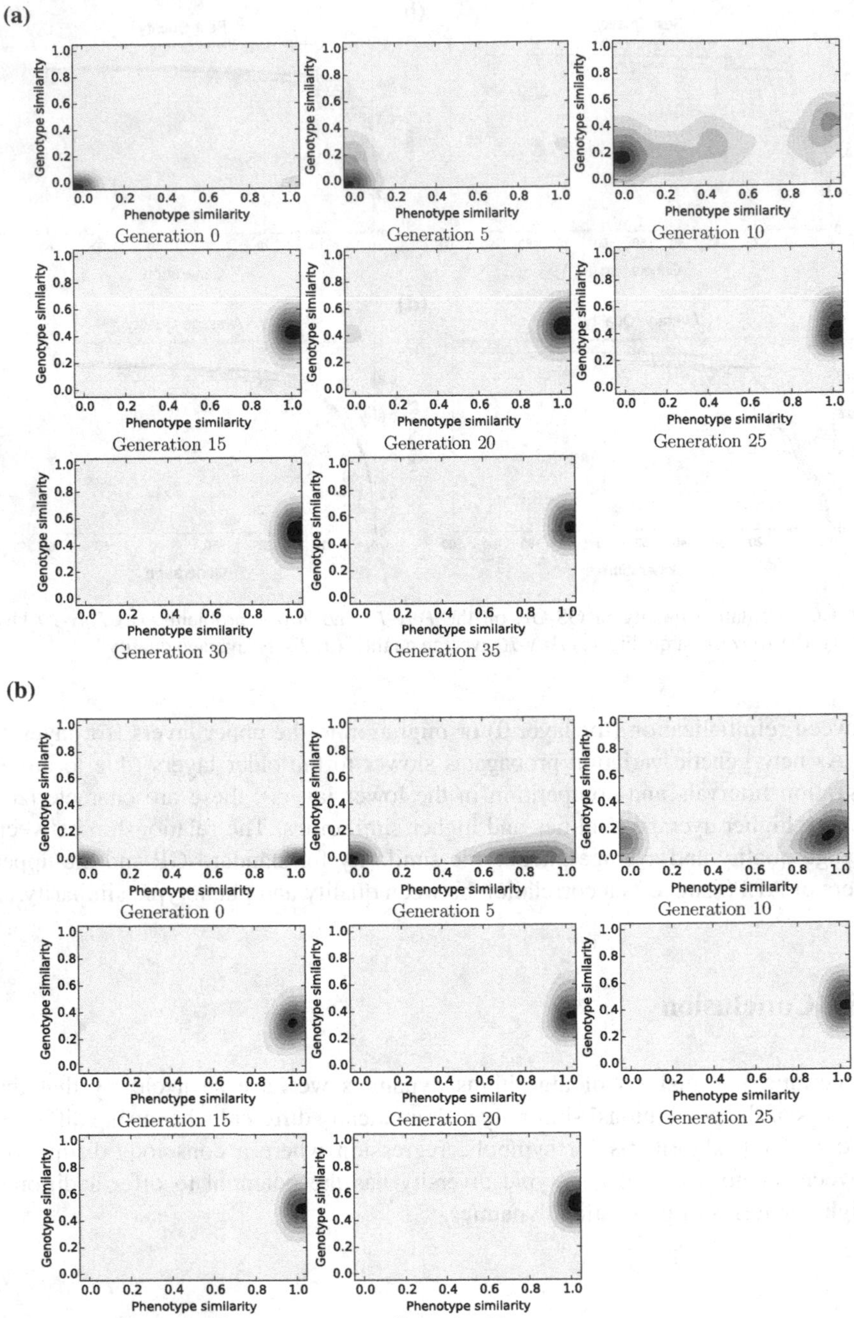

Fig. 1.7 Genotypic vs. phenotypic similarities in OS-GP. (**a**) *Poly-10* problem. (**b**) *Tower* problem

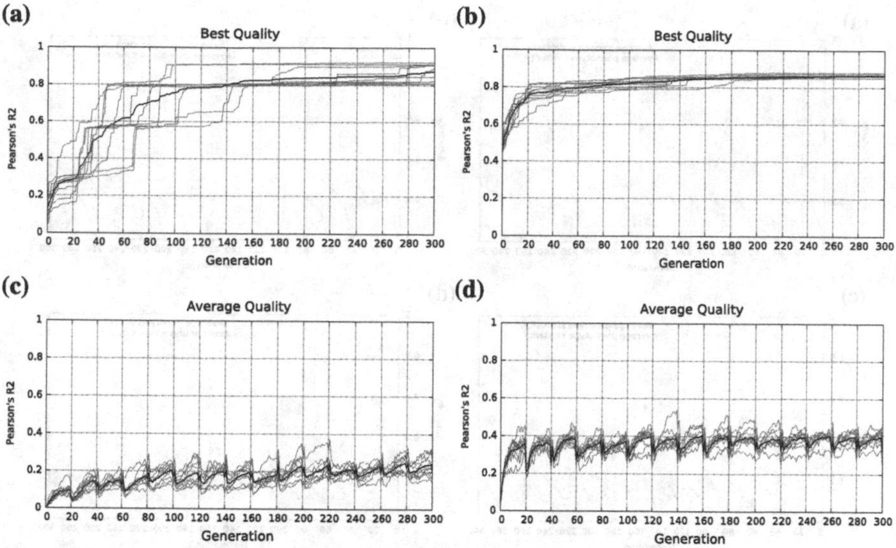

Fig. 1.8 Best and average qualities in ALPS-GP solving the problem instances *Poly-10* and *Tower*. (**a**) *Poly-10* best quality. (**b**) *Tower* best quality. (**c**) *Poly-10* average quality. (**d**) *Tower* average quality

In future work we will research in more detail on the effects of the choice of parent selection mechanisms, varying settings for offspring selection and parent selection pressure, and varying aging parameter settings of ALPS-GP.

However, apart from the considered aspect of analyzing the dynamic characteristics of different algorithm flavors a detailed analysis of genotypic and phenotypic similarity and distance measures should also be helpful for the selection of models for ensemble approaches in symbolic regression. In order to fulfill the general claim of ensemble based modeling it is important to use high quality independent models for ensemble interpretation.

In this context it should be helpful to perform the model selection task in a way that good (with respect to training fitness) models with high phenotypic as well as genotypic diversity are selected out of different symbolic regression/classification runs using different algorithm flavors, different parameter settings, as well as different functional bases.

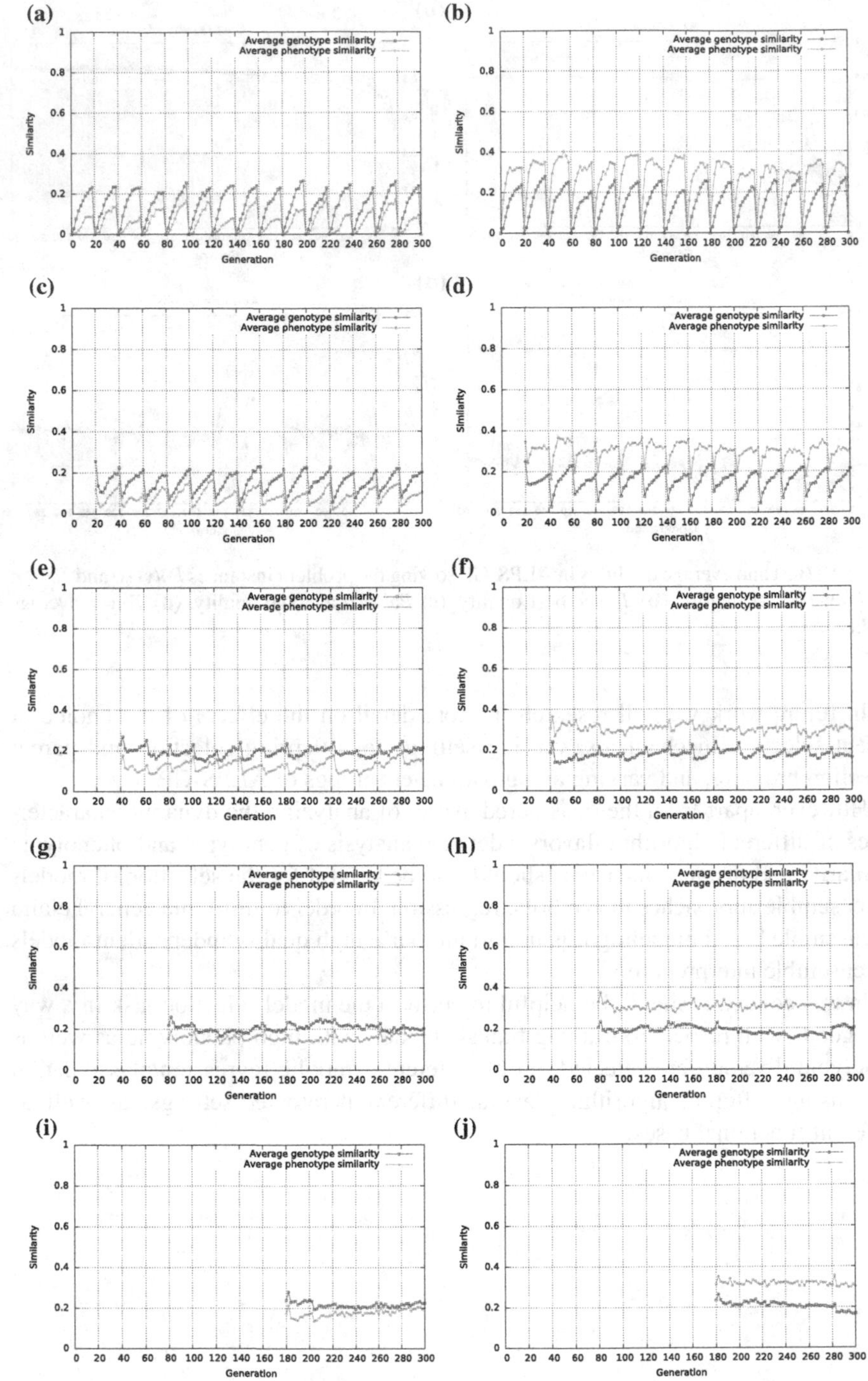

Fig. 1.9 Genotypic and phenotypic similarity in ALPS-GP solving the problem instances *Poly-10* and *Tower*. (**a**) *Poly-10*, layer 0. (**b**) *Tower*, layer 0. (**c**) *Poly-10*, layer 1. (**d**) *Tower*, layer 1. (**e**) *Poly-10*, layer 2. (**f**) *Tower*, layer 2. (**g**) *Poly-10*, layer 3. (**h**) *Tower*, layer 3. (**i**) *Poly-10*, layer 4. (**j**) *Tower*, layer 4

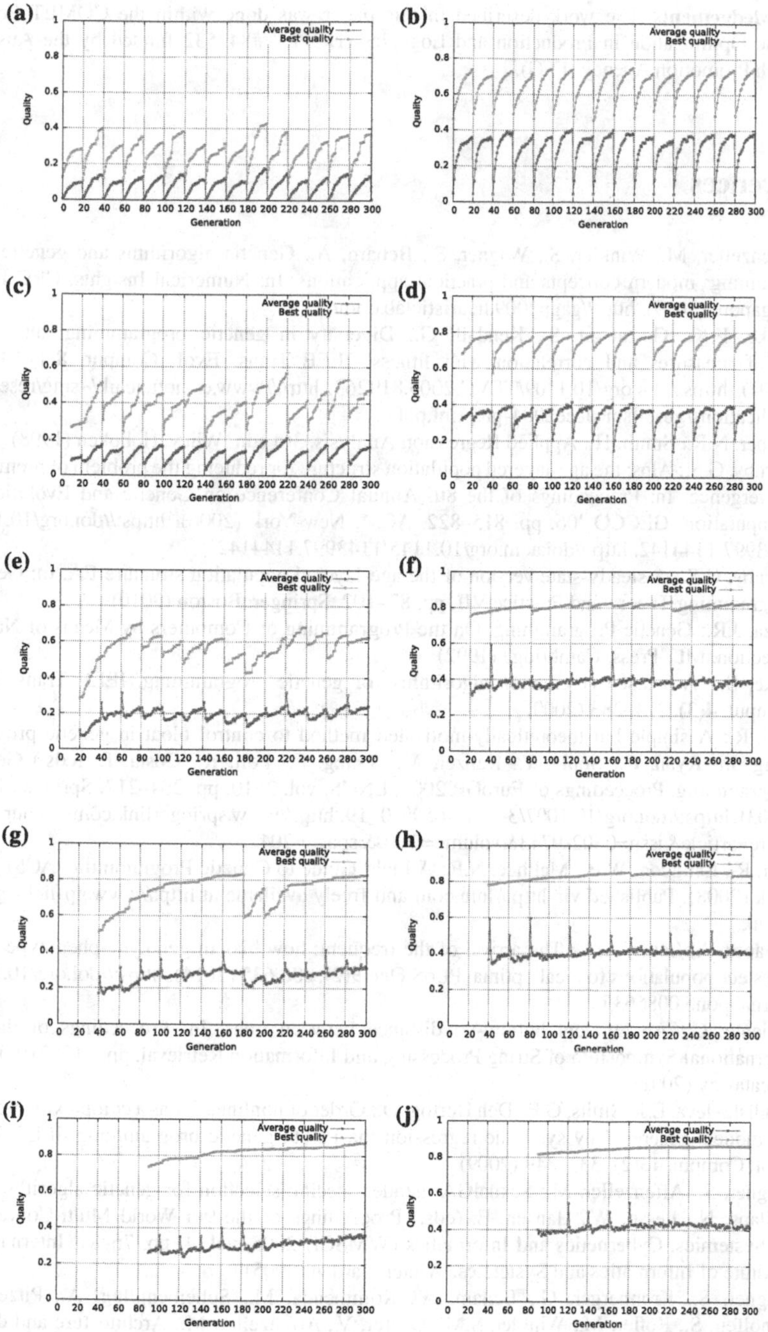

Fig. 1.10 Average quality per layer in ALPS-GP solving the problem instances *Poly-10* and *Tower*. **(a)** *Poly-10*, layer 0. **(b)** *Tower*, layer 0. **(c)** *Poly-10*, layer 1. **(d)** *Tower*, layer 1. **(e)** *Poly-10*, layer 2. **(f)** *Tower*, layer 2. **(g)** *Poly-10*, layer 3. **(h)** *Tower*, layer 3. **(i)** *Poly-10*, layer 4. **(j)** *Tower*, layer 4

Acknowledgements The work described in this paper was done within the COMET Project Heuristic Optimization in Production and Logistics (HOPL), #843532 funded by the Austrian Research Promotion Agency (FFG).

References

1. Affenzeller, M., Winkler, S., Wagner, S., Beham, A.: Genetic algorithms and genetic programming: modern concepts and practical applications. In: Numerical Insights. CRC Press, Singapore (2009). http://gagp2009.heuristiclab.com/
2. Burke, E.K., Gustafson, S., Kendall, G.: Diversity in genetic programming: an analysis of measures and correlation with fitness. IEEE Trans. Evol. Comput. **8**(1), 47–62 (2004). https://doi.org/10.1109/TEVC.2003.819263. http://www.cs.nott.ac.uk/~smg/research/publications/gustafson-ieee2004-preprint.pdf
3. Draper, N.R., Smith, H.: Applied Regression Analysis, 3rd edn. Wiley, Hoboken (1998)
4. Hornby, G.S.: Alps: the age-layered population structure for reducing the problem of premature convergence. In: Proceedings of the 8th Annual Conference on Genetic and Evolutionary Computation, GECCO '06, pp. 815–822. ACM, New York (2006). https://doi.org/10.1145/1143997.1144142. http://doi.acm.org/10.1145/1143997.1144142
5. Hornby, G.S.: A steady-state version of the age-layered population structure EA. In: Genetic Programming Theory and Practice VII, pp. 87–102. Springer, Boston (2010)
6. Koza, J.R.: Genetic Programming: On the Programming of Computers by Means of Natural Selection. MIT Press, Cambridge (1992)
7. Luke, S.: Two fast tree-creation algorithms for genetic programming. IEEE Trans. Evol. Comput. **4**(3), 274–283 (2000)
8. Poli, R.: A simple but theoretically-motivated method to control bloat in genetic programming. In: Ryan, C., Soule, T., Keijzer, M., Tsang, E., Poli, R., Costa, E. (eds.) Genetic Programming, Proceedings of EuroGP'2003, LNCS, vol. 2610, pp. 204–217. Springer, Essex (2003). https://doi.org/10.1007/3-540-36599-0_19. http://www.springerlink.com/openurl.asp?genre=article&issn=0302-9743&volume=2610&spage=204
9. Poli, R., Langdon, W.B., McPhee, N.F.: A Field Guide to Genetic Programming. ACM, New York (2008). Published via http://lulu.com and freely available at http://www.gp-field-guide.org.uk
10. Schaper, S., Louis, A.A.: The arrival of the frequent: how bias in genotype-phenotype maps can steer populations to local optima. PLoS One **9**(2), e86,635 (2014). https://doi.org/10.1371/journal.pone.0086635
11. Valiente, G.: An efficient bottom-up distance between trees. In: Proceedings of the 8th International Symposium of String Processing and Information Retrieval, pp. 212–219. IEEE, Piscataway (2001)
12. Vladislavleva, E.J., Smits, G.F., Den Hertog, D.: Order of nonlinearity as a complexity measure for models generated by symbolic regression via Pareto genetic programming. IEEE Trans. Evol. Comput. **13**(2), 333–349 (2009)
13. Wagner, S., Affenzeller, M.: SexualGA: gender-specific selection for genetic algorithms. In: Callaos, N., Lesso, W., Hansen, E. (eds.) Proceedings of the 9th World Multi-Conference on Systemics, Cybernetics and Informatics (WMSCI) 2005, vol. 4, pp. 76–81. International Institute of Informatics and Systemics, Winter Garden (2005)
14. Wagner, S., Kronberger, G., Beham, A., Kommenda, M., Scheibenpflug, A., Pitzer, E., Vonolfen, S., Kofler, M., Winkler, S.M., Dorfer, V., Affenzeller, M.: Architecture and design of the heuristiclab optimization environment. Adv. Methods Appl. Comput. Intell. Top. Intell. Eng. Inform. **6**, 197–261 (2013)

15. White, D.R., McDermott, J., Castelli, M., Manzoni, L., Goldman, B.W., Kronberger, G., Jaskowski, W., O'Reilly, U.M., Luke, S.: Better GP benchmarks: community survey results and proposals. Genet. Program Evolvable Mach. **14**(1), 3–29 (2013). https://doi.org/10.1007/s10710-012-9177-2. http://gpbenchmarks.org/wp-content/uploads/2014/09/GP-Benchmarks-GPEM-2013-preprint-correction-v2.pdf
16. Winkler, S.M.: Structural versus evaluation based solutions similarity in genetic programming based system identification. In: González, J.R., Pelta, D.A., Cruz, C., Terrazas, G., Krasnogor, N. (eds.) Nature Inspired Cooperative Strategies for Optimization, NICSO 2010. Studies in Computational Intelligence, vol. 284, pp. 269–282. Springer, Granada (2010). https://doi.org/10.1007/978-3-642-12538-6_23

Chapter 2
An Investigation of Hybrid Structural and Behavioral Diversity Methods in Genetic Programming

Armand R. Burks and William F. Punch

Abstract Premature convergence is a serious problem that plagues genetic programming, stifling its search performance. Several genetic diversity maintenance techniques have been proposed for combating premature convergence and improving search efficiency in genetic programming. Recent research has shown that while genetic diversity is important, focusing directly on sustaining behavioral diversity may be more beneficial. These two related areas have received a lot of attention, yet they have often been developed independently. We investigated the feasibility of hybrid genetic and behavioral diversity techniques on a suite of problems.

Keywords Genetic programming · Premature convergence · Genetic diversity · Structural diversity · Behavioral diversity · Semantics

2.1 Introduction

Premature convergence is a well-known problem, in not only genetic programming (GP) but also many evolutionary algorithms (EAs). We refer to premature convergence as the phenomenon that occurs during the evolutionary process when the population becomes largely homogeneous, either genetically or phenotypically, causing poor, inefficient exploration of the search space. This often occurs early in the evolutionary process and dramatically decreases the probability of discovering a solution.

Premature convergence in GP is perhaps even more perplexing than in other EAs. Because genotypes are represented by a variable-length tree structure in classical GP, convergence of GP populations is much different than in fixed-length representations, for example a binary-encoded genetic algorithm (GA). When

A. R. Burks (✉) · W. F. Punch
BEACON Center for the Study of Evolution in Action, Michigan State University, East Lansing, MI, USA
e-mail: burksarm@msu.edu; punch@msu.edu

R. Riolo et al. (eds.), *Genetic Programming Theory and Practice XIV*, Genetic and Evolutionary Computation,
https://doi.org/10.1007/978-3-319-97088-2_2

single-point crossover is performed on two identical parents in a binary-encoded GA, the resulting offspring will also be identical to the parents. However, since GP trees typically encode functions, crossover of two identical trees can still create functionally different offspring because the subtrees exchanged between parents will likely be different. Therefore, even in a largely genetically homogeneous GP population, it is still possible for different phenotypes to evolve.

Although GP populations do not converge in the same way as other EAs, it has been shown that GP populations often experience a very rapid loss of diversity near the onset of the evolutionary process [19]. Within a very few generations, a large percentage of the population (over 70%) became identical to a single first-generation ancestor in the top four levels of the trees [19]. This is important because the rooted structures define the functional context of the programs, and changes near the root have a more dramatic impact on the overall functionality of programs. Further analysis in that same study showed that at the final generation, the population contained genetic material from very few individuals from the first generation.

Premature convergence can also be observed in the phenotypic (behavioral) space in GP [12, 13, 15]. A complication with measuring diversity in the genotypic space, especially in tree-based GP, is that of the complex genotype-to-phenotype mapping. Since many trees, although structurally and genetically different, can evaluate to the same expression or behavior, a genetically diverse population does not necessarily guarantee behavioral diversity. Conversely, a genetically converged GP population is not necessarily behaviorally converged. However, it has been shown that GP populations can suffer poor behavioral diversity [12, 13], becoming concentrated to a few clusters of behaviors even though the individuals may be genetically different [15].

Fortunately, a lot of effort has gone into understanding premature convergence and developing techniques to overcome it. Many approaches focus on preserving genetic diversity in the population as a means of avoiding premature convergence [5, 9, 25]. Depending on how genetic diversity is defined and how it is enforced, the success of this type of approach can vary greatly [2]. In addition to the genetic diversity methods, more recent research has focused on explicitly operating in the behavioral space, overcoming the issues with the genotype-to-phenotype mapping.

2.2 Related Work

We briefly discuss some of the related approaches that have been used to address premature convergence in GP. While this section does not fully survey such approaches, we discuss some relevant approaches and those used in this study.

2.2.1 Genetic Diversity Techniques

Many genetic diversity techniques have been developed for avoiding premature convergence in GP. Some of these approaches are based on well-known techniques from the broader EA and GA literature, such as crowding [4] and fitness sharing [7]. As a result, these techniques are more general and can be more widely applied [9, 11, 25], while other approaches are more closely tied to GP [3, 5, 10].

Age-Fitness Pareto Optimization [25], which we will refer to as A/F for brevity, uses *genotypic age* to preserve genetic diversity. Genotypic age is a measure of how long the oldest portion of an individual's genotype has been evolving in the population. A/F uses a multi-objective approach to simultaneously optimize fitness while minimizing genotypic age. This allows new genetic material to propagate through the population and perhaps explore different regions of the search space. A benefit of this approach is that it removes the extra overhead of managing a more complicated population structure such as in [9], while it was shown to achieve similar or better results [25].

Instead of genotypic age, the genetic marker diversity algorithm (GMD-GP) focuses on preventing the topmost portions of the trees from converging to the same structure [3]. This is done by using the top fragments of the trees as *genetic markers*. Genetic markers are created by beginning at a certain level in the tree and performing a depth-first traversal to a certain depth. The genetic marker is then a partial Lisp-style expression containing all the visited nodes.

As a measure of structural diversity, GMD-GP calculates the density of each genetic marker, which is the fraction of individuals in the population that contain the genetic marker. GMD-GP then uses the density of each genetic marker to prevent a single structure from becoming too widespread by minimizing density as a second objective while simultaneously optimizing fitness in a multi-objective optimization scheme. This promotes the search for new structures while still maintaining global selection pressure to improve fitness. GMD-GP was shown to find solutions significantly faster than A/F in different problem domains [3].

2.2.2 Semantic Methods in GP

GP has largely been applied to problems containing several input cases, Y_i, for each of which there is a corresponding output, X_i, and fitness is often based on each input-output pair. The semantics or behavior of a GP individual has therefore been defined as the vector of output values obtained by executing a program on a set of input cases [20]. We will use the terms semantics and behavior interchangeably to refer to the output vector of an individual. The behavioral or semantic space then refers to the space of all such output vectors.

A number of semantic-aware crossover operators have been introduced, ranging from rejecting offspring that are semantically equivalent to their parents [1, 22]

Algorithm 1 Lex-C: count bias sorting of fitness cases

Require: cases ▷ Set of fitness cases for the given problem.
$counts \Leftarrow \emptyset$ ▷ Map of total individuals that solve each fitness case.
for $case \in cases$ **do**
  $counts[case] \Leftarrow$ total individuals that solve $case$
end for
return $sortAscending(counts)$ ▷ Return the sorted fitness case indices in ascending order.

to considering the semantic similarity of the subtrees to be exchanged between parents [27]. The geometric semantic crossover operator forces the semantics of an offspring to lie between those of its parents in the semantic space [21]. This way, the offspring is no worse than its least-fit parent, in terms of fitness. A recent extension of this technique addressed the issue of exponential growth of offspring in geometric semantic crossover by exchanging a subtree that most closely resembles the semantics of an individual, rather than using the entire tree [23].

Semantic methods that operate during the selection phase have also been proposed [8, 22]. Semantic sharing was introduced as an extension of fitness sharing that considers the semantic, rather than syntactic, similarity of trees [22]. To calculate the distance between individuals, Sampling Semantics Distance is used to measure the difference between individuals' output vectors. This avoids the expensive calculation of the syntactic distance of trees, and it is immune to the genotype-to-phenotype mapping issue.

2.3 Fitness Case Bias in Lexicase Selection

Lexicase Selection is a technique that imposes semantic diversity on the population during the selection phase [8]. Lexicase Selection was inspired by the fact that the most common class of problems to which GP has been applied requires a solution that must perform well on several fitness cases rather than just one. Lexicase Selection focuses on selecting parents that perform well on different sets of fitness cases, with the goal that offspring inherit complementary features from the parents with respect to the fitness cases.

Lexicase Selection proceeds as follows. The entire population is first set as the pool of candidates for selection. Next, the list of fitness cases to be used is the whole set of fitness cases for the problem. The order of these cases is then randomized. The main loop begins with the first fitness case in the list and updates the candidates to contain only the individuals that have exactly the best error on that fitness case. Then, that fitness case is removed and the loop continues until either (1) only one candidate remains or (2) all fitness cases have been considered. In the latter condition, a randomly selected remaining candidate is returned.

We examined two extensions of Lexicase Selection that address the question of whether placing bias on certain fitness cases yields any advantage compared to

Algorithm 2 Lex-E: error bias sorting of fitness cases

Require: cases ▷ Set of fitness cases for the given problem.
errors ⇐ ∅ ▷ Map of average error for each fitness case.
for *case* ∈ *cases* **do**
    errors[*case*] ⇐ average population error on *case*
end for
return *sortDescending*(*errors*) ▷ Return the sorted fitness case indices in descending order.

random sorting of fitness cases. The rationale behind this is based on the observation that over time some fitness cases are solved by a large percentage of the population, while some fitness cases are more rarely solved or individuals tend to commit higher error on them. Placing bias towards fitness cases that are currently more difficult may allow the problem to be solved faster. Related techniques such as implicit fitness sharing [18] and Keep-Worst Interleaved Sampling (KW-IS) [16] have been proposed to bias selection towards different fitness cases over time. Implicit fitness sharing reduces the fitness contribution of each fitness case based on the total individuals that solve the fitness case. KW-IS alternates between using all the fitness cases for calculating fitness in some generations and using some number of the current most difficult fitness cases in other generations.

The first extension to Lexicase Selection (referred to as Lexicase Selection with count bias, Lex-C), shown in Algorithm 1, orders the fitness cases based on the total number of individuals that solve each fitness case, where the fitness case solved by the fewest individuals is first in the list. This way, bias is placed on the current most difficult of the fitness cases that are solved by at least some number of individuals in the population. It is possible that the most difficult fitness case is not solved by any individual; in this situation, Lex-C still favors individuals that performed best on that test case. Since this ordering of the fitness cases is based on the current population, the algorithm would always select the same individual if the entire population were used as candidates (as in the original algorithm). To avoid this, we instead select a random subset of the individuals.

The second extension we considered (referred to as Lexicase Selection with error bias, Lex-E), described in Algorithm 2, sorts the fitness cases in descending order by average error committed by the population. This way, the current most difficult fitness case is used in each iteration to retain only the candidates with the exact best error on that fitness case. This differs from count bias in that the fitness case with the highest average error may not be solved by any individual in the current population. However, the algorithm still selects from the individuals that performed the best on that fitness case. For discrete problems, in which the test condition is pass or fail such as in Boolean problems, Lex-E behaves identically to Lex-C. As in Lex-C, we also must use a subset of the population as candidates, since the ordering of the fitness cases does not change during the current generation.

2.4 Hybrid Structural and Behavioral Diversity Methods

Although several genetic and behavioral diversity techniques have been proposed as a means of improving the efficiency of GP, these two areas have often been developed independently. While genetic and behavioral diversity are very closely connected and undoubtedly affect each other, we are interested in whether or not hybrid methods that simultaneously focus on both provide any advantage over each approach in isolation, as well as how they compare to a traditional GP setup.

Hybrid approaches that explicitly focus on genetic and behavioral diversity are of interest because they may address some of the issues experienced by both techniques. As we discussed in Sect. 2.1, genetic diversity does not necessarily imply behavioral diversity because the same behavior can be expressed by different genotypes. Therefore, focusing explicitly on genetic diversity may not be as beneficial as focusing directly on behavioral diversity. Additionally, previous studies have shown that the widely used genetic operators in GP can lead to convergence in the genotypic space [6, 19, 24]. This means that while semantic diversity selection techniques focus on maintaining a diverse set of behaviors in the population, they may still benefit from genetic diversity.

We investigated hybridizing GMD-GP with the different versions of Lexicase Selection that we discussed in Sect. 2.3. In a previous study, GMD-GP was shown to outperform other genetic diversity algorithms in different problem domains [3]. However, since the multi-objective approach imposes selection pressure after breeding, GMD-GP uses random mating to select parents. Therefore, a natural extension for hybridizing GMD-GP is to incorporate behavioral diversity into the selection phase. In the experiments discussed below, we compared the following hybrid approaches: GMD-GP with Lexicase Selection (abbreviated as GMD+Lex): GMD-GP with Lexicase Selection and count bias (GMD+Lex-C), and GMD-GP with Lexicase Selection and error bias (GMD+Lex-E).

2.5 Experimental Setup

The experiments we conducted are two-fold. First, we compared the performance of each algorithm based on how often it solved the problems in the test suite. Next, we compared the effects each approach has on different measures of diversity over time. For the test suite, we used several problems adapted from previous work [14, 17]. Because we used Lexicase Selection, these problems are all error-based and contain multiple fitness cases. The problems fall into one of two domains: symbolic regression (16 problems), and finite algebras (10 problems).

Table 2.1 General GP settings common to all problems

Parameter	Value
Random initialization	Ramped half & half
Maximum evaluations	1 million
Total trials	100
Population size	500
Tournament size	2
Max. tree size	512
Max. tree depth	17
Crossover probability	0.90
Reproduction probability	0.10
Elites	50 (10%)
GMD-GP genetic marker depth	2

Table 2.1 lists the general settings. For all experiments, we used the GP framework upon which GMD-GP was developed.[1] For genetic markers in GMD-GP, we used the top three tree levels, as these settings were the best in preliminary experiments. Since GMD-GP uses binary tournaments in its selection phase [3], we also used binary tournament selection in the standard GP setup. We used the common tournament size of seven for the Lexicase Selection extensions we described in Sect. 2.3, since the whole population can not be used. Although we did not tune this parameter, additional analysis into how the tournament size affects the performance of Lex-E and Lex-C may provide additional insight into these extensions, and we intend to investigate this in the future.

2.5.1 Problems

The first class of problems consisted of several one- and two-variable symbolic regression problems that have been used in previous work such as [14], and recommended as potential benchmark problems for testing GP [17]. The general task for these problems is the same: using a set of training points (i.e., fitness cases) within a given domain, evolve a mathematical expression that most closely matches a target function. Fitness is therefore defined as a function of the cumulative absolute error across each fitness case. For the training points, we used the previously-defined ranges specified by McDermott et al. [17], with the corrected values that were later given.[2]

The function and terminal set for each problem is the same, the only difference being the number of variables, x_i, used in the terminal set corresponding to the number of variables required for the particular problem. For the function set, we

[1] https://github.com/burks-pub/gecco2015.

[2] https://cs.gmu.edu/~sean/papers/gecco12benchmarks3.pdf.

used the following, F = {+, −, *, %, SIN, COS, LOG, EXP, $-x$ }, where % represents protected division and LOG represents the protected natural logarithm operator. As in previous work [14], the terminal set did not include random constants.

The second class of problems was used in a previous study in which GP was applied (with human-competitive results) to finite algebras [26]. These problems include a single binary operator that operates within the ternary domain and only has defined outputs for the input values {0, 1, 2}. The operator contains an underlying algebra that defines its input-output mapping. There are two tasks that we considered, and for each task, we used the operators $A1$–$A5$, which are defined in [26], giving a total of ten problems. The terminal set consisted of a terminal node corresponding to each inlineput. The function set included the single operator A_i for the corresponding algebra being used. Following [26], fitness was calculated as a function of the total error on the fitness cases.

The first task involves using the single operator to evolve an expression that encodes the *discriminator term*, which returns x if $x \neq y$, and z otherwise, for inputs x, y, z. The fitness cases are then all combinations of the three inputs, x, y, z, resulting in a total of $3^3 = 27$ fitness cases. The second task involves evolving an expression that encodes a *Mal'cev term* that must satisfy the following: $m(x, x, y) \approx m(y, x, x) \approx y$. Due to this definition, not all possible combinations of the inputs have a defined output, resulting in a total of 15 fitness cases.

2.6 Results

Due to the large number of problems in the test suite, and the number of methods we compared, the reported results are aggregated across all problems. The figures below show pairwise comparisons of the percentage of problems in which the method shown above the plot performed significantly better than each other method, for the given metric under consideration. The percentages for any pair of methods need not sum to 100%, because the differences between certain methods were not statistically significant for every problem. To determine statistical significance, we used a pairwise Fisher's exact test for comparing the success rates, and a Mann–Whitney U test for all the other metrics.

In order to determine whether or not the hybrid structural and behavioral diversity methods described in Sect. 2.4 perform better than either method alone, as well as how they compare to the standard GP setup, we compared the **success rate**: the percentage of trials in which each method found a solution within the computational budget of 1 million fitness evaluations. We refer to standard GP as the classic generational GP using tournament selection and random subtree crossover with 90% internal node bias.

The pairwise comparison of the success rates is shown in Fig. 2.1. The GMD plot shows that GMD-GP outperformed standard GP and Lex in a majority of the symbolic regression problems. The plots for GMD+Lex-E and GMD+Lex-

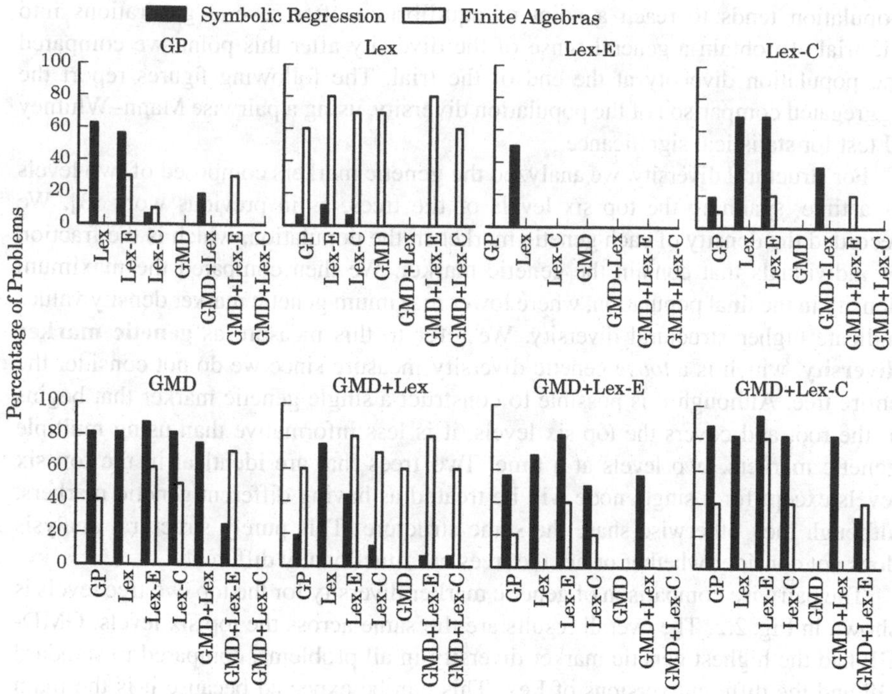

Fig. 2.1 Success rate pairwise comparison: percentage of problems in which the method shown above each plot had a significantly higher success rate than each method listed

C show that they both had significantly higher success rates than both standard GP and Lex in a majority of the symbolic regression problems, while all three hybrids significantly improved the success rate of Lex. Likewise, both Lex-E and (especially) Lex-C improved the success rate of Lex in a majority of the symbolic regression problems.

Figure 2.1 also shows that, with the exception of the GMD+Lex hybrid, Lex had a significantly higher success rate than every other method, including GMD-GP, in most of the finite algebras problems. On the other hand, the GMD+Lex hybrid significantly improved the success rate of GMD-GP in around 60% of the finite algebras problems. This demonstrates the utility of such hybridization; Lex tended to perform poorly for symbolic regression but the hybrids significantly improved upon Lex, and likewise GMD-GP performed significantly worse than Lex for the finite algebras problems but the GMD+Lex hybrid significantly improved upon GMD-GP and had identical success rates to Lex, which outperformed all the other methods.

We also compared the methods using different measures of diversity in a second set of experiments. For each problem, we allowed each algorithm to continue for 250,000 fitness evaluations, and we analyzed the trend of the different diversity metrics we discuss below. The general trend for the diversity metrics is that the population tends to reach a point of equilibrium after several generations into the trial. To obtain a general sense of the diversity after this point, we compared the population diversity at the end of the trial. The following figures report the aggregated comparison of the population diversity, using a pairwise Mann–Whitney U test for statistical significance.

For structural diversity, we analyzed the genetic markers composed of two levels at a time, spanning the top six levels of the trees, as in previous work [3]. We recorded the density of each genetic marker in the population, which is the fraction of individuals that contain the genetic marker. We then compared the maximum density in the final population, where lower maximum genetic marker density values indicate higher structural diversity. We refer to this measure as **genetic marker diversity**, which is a *loose* genetic diversity measure since we do not consider the entire tree. Although it is possible to construct a single genetic marker that begins at the root and covers the top six levels, it is less informative than using multiple genetic markers, two levels at a time. Two trees that are identical in the top six levels except for a single node will be treated as having different genetic markers, although they otherwise share the same structure. This purely structural analysis does not consider whether or not the trees are functionally different.

The pairwise comparison of genetic marker diversity for the top two tree levels is shown in Fig. 2.2. The overall results are the same across the top six levels. GMD-GP had the highest genetic marker diversity in all problems, compared to standard GP and the different versions of Lex. This can be expected because it is the main focus of GMD-GP. Since GMD-GP already maintains very high genetic marker diversity, it was generally decreased by the hybridization, although the hybrids increased the genetic marker diversity of GMD-GP in some cases in the finite algebras problems.

While Lex had significantly higher diversity than standard GP in the symbolic regression problems, it still tended to reach high genetic marker density values (i.e., relatively low genetic marker diversity), near 50–60% in the topmost portions of the trees. The hybridization of GMD-GP and Lex significantly increased the genetic marker diversity of Lex (shown in the Lex bar of each of the plots for the hybrids). These results suggest that this increased structural diversity leads to the improved performance over Lex that we observed in Fig. 2.1 for the symbolic regression problems.

Apart from genetic marker diversity, we compared the behavioral diversity of the different methods, using the definition of behavior based on the output vectors of the individuals in the population [13]. Two individuals are behaviorally different if their output vectors differ by at least one element. Behavioral diversity is then defined as the fraction of unique behaviors in the population. In the symbolic regression problems (unlike the finite algebras problems, wherein the possible outputs are integers), we used a difference threshold of 0.01 for the real-valued outputs.

Figure 2.3 shows the pairwise comparison of the behavioral diversity of the final populations across all the problems. While GMD-GP maintains a high level of structural diversity, it experiences less behavioral diversity than that of standard GP and Lex in many of the problems (Lex often had lower behavioral diversity

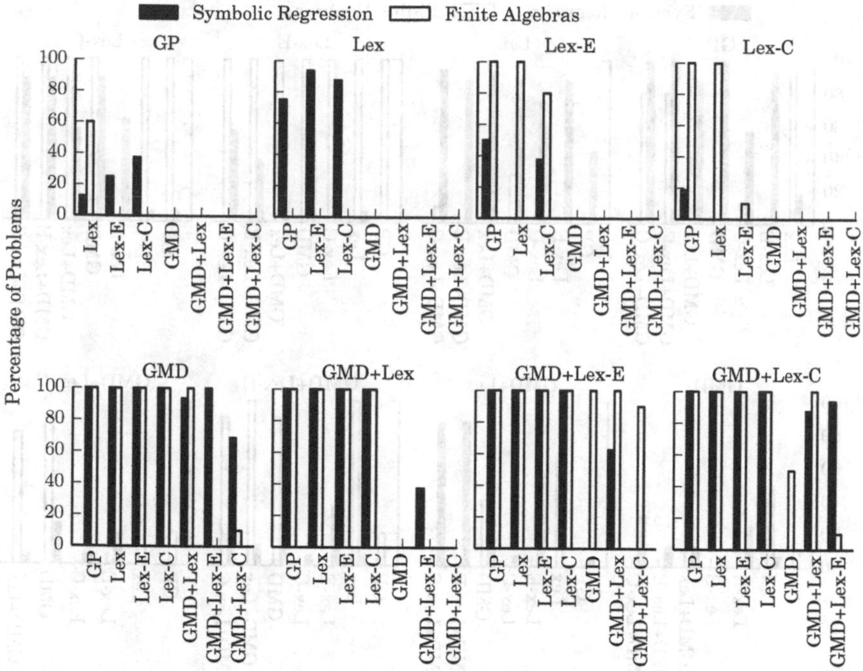

Fig. 2.2 Genetic marker diversity pairwise comparison: percentage of problems in which the final population of the method shown above each plot had significantly higher ending genetic marker diversity than each method listed

than GMD-GP in the finite algebras problems because Lex typically converged to the solution very quickly although the trial always lasted for 250,000 fitness evaluations). On the other hand, the hybridization significantly increased the behavioral diversity of GMD-GP in several of the problems (shown in the GMD bar of each of the bottom three plots in Fig. 2.3). This suggests that the hybridization with Lex allows GMD-GP to explore more behaviors through its genetically diverse population, whereas GMD-GP alone often discovers the same behaviors through different structures because of the high genetic diversity.

Lex maintained significantly higher behavioral diversity than standard GP and GMD-GP in a large number of the symbolic regression problems. While the hybridization decreased the behavioral diversity of Lex in the symbolic regression problems, the hybrids increased the behavioral diversity of Lex in the finite algebras problems. However, this increased behavioral diversity did not yield performance improvements over Lex in the finite algebras problems, as we observed in Fig. 2.1. This may be due to the fact that the random subtree crossover operator does not make efficient use of the added diversity, and we are investigating this further, as we discuss in Sect. 2.7.

Finally, we compared the standard deviation of the fitness values in the final populations of each method. This is more informative than comparing the total

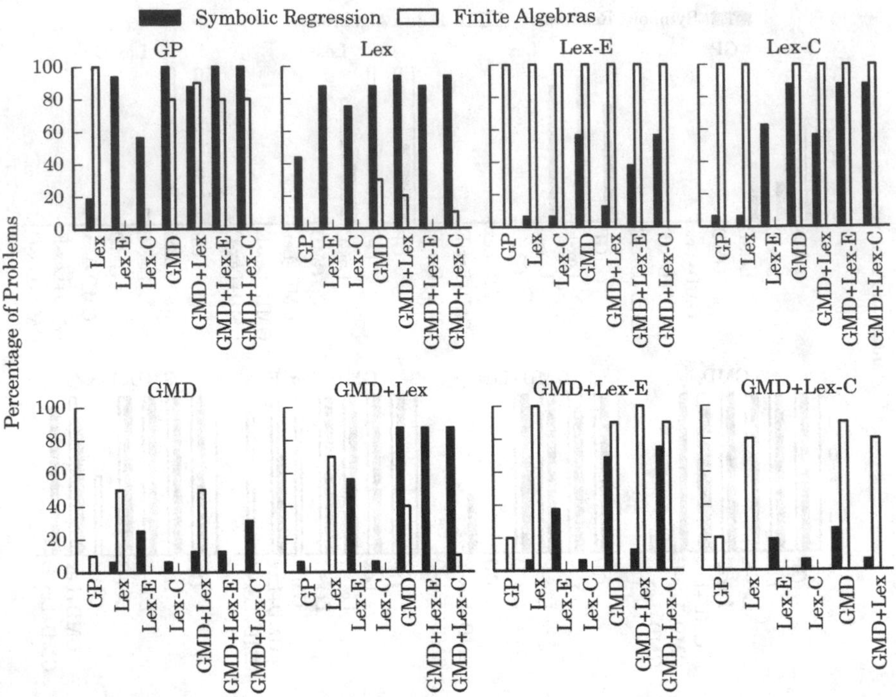

Fig. 2.3 Behavioral diversity pairwise comparison: percentage of problems in which the final population of the method shown above the plot had significantly higher behavioral diversity than that of each method listed

unique fitness values alone, as it considers the difference between the fitness values in the population, which is an indication of the spread in the fitness landscape. Figure 2.4 shows that in many cases the hybrid methods increased the fitness diversity over GMD-GP and Lex in several problems. Related to the problem of genotype-to-phenotype mapping that we discussed in Sect. 2.1, different behaviors can result in the same fitness value, so a behaviorally diverse population does not necessarily guarantee effective exploration of the fitness landscape. Additionally, a population with more unique fitness values does not guarantee a good spread across the range of possible fitness values. This is evidenced by the fact that while GMD-GP populations had lower behavioral diversity than Lex and standard GP in the symbolic regression problems, the standard deviation in fitness was significantly higher in GMD-GP in all of the symbolic regression problems (see the GP and Lex bars of the GMD plot in Fig. 2.4). This increased standard deviation in fitness is then gained by Lex through the hybridization with GMD-GP. The same result holds for GMD-GP versus GMD+Lex in the finite algebras problems.

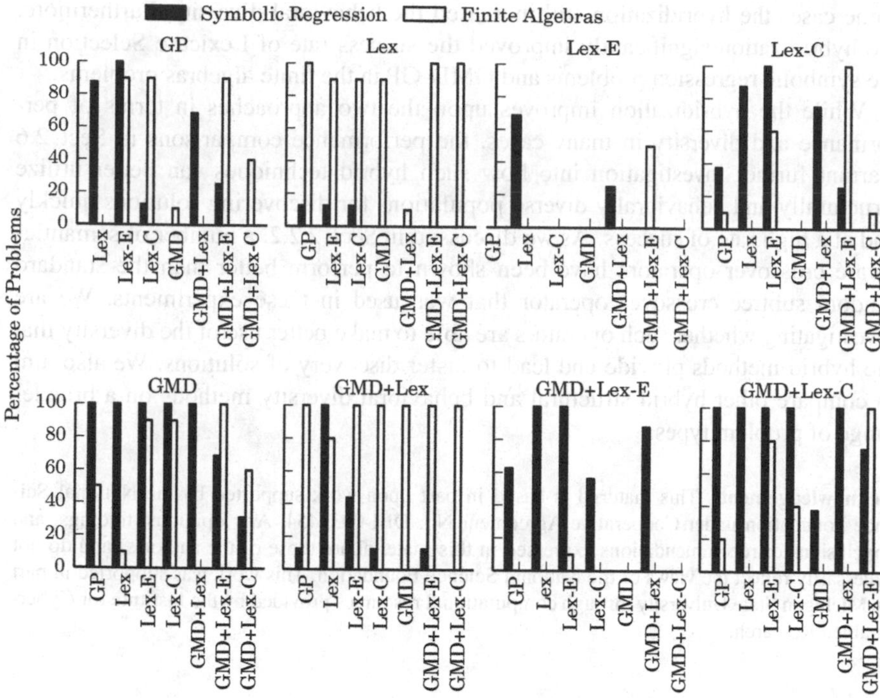

Fig. 2.4 Fitness standard deviation pairwise comparison: percentage of problems in which the final population of the method shown above the plot had significantly higher standard deviation of fitness than that of each method listed

2.7 Conclusions and Future Work

We have investigated hybridizing two techniques that focus on maintaining diversity in GP populations through two separate means: structural and behavioral diversity. The goal of such hybridization is that, by simultaneously focusing on both structural and behavioral diversity, the hybrid methods can address the shortcomings of both approaches and achieve better search performance and efficiency than either approach in isolation. While the hybrid techniques did not dominate both approaches across all the problems in the test suite, the experimental results show promise for the utility of such hybrid approaches.

Specifically, GMD-GP often experienced lower behavioral diversity than the other approaches while maintaining significantly higher structural diversity. The hybridization with Lexicase Selection increased the behavioral diversity of GMD-GP, with the trade-off of less (yet still high) structural diversity. Similarly, Lexicase Selection often had lower structural diversity than GMD-GP but higher behavioral diversity, while the hybridization significantly increased the structural diversity and fitness diversity of Lexicase Selection. The trade-off in this case was lower behavioral diversity compared to Lexicase Selection alone in many cases, while in some cases the hybridization still increased the behavioral diversity. Furthermore, the hybridization significantly improved the success rate of Lexicase Selection in the symbolic regression problems and GMD-GP in the finite algebras problems.

While the hybridization improves upon the two approaches in terms of performance and diversity in many cases, the performance comparisons in Sect. 2.6 warrant further investigation into how such hybrid techniques can better utilize structurally and behaviorally diverse populations for discovering solutions quickly and at a high rate of success. As we discussed in Sect. 2.2.2, a number of semantic-aware crossover operators have been shown to perform better than the standard random subtree crossover operator that was used in these experiments. We are investigating whether such operators are able to make better use of the diversity that the hybrid methods provide and lead to faster discovery of solutions. We also aim to compare other hybrid structural and behavioral diversity methods on a broader range of problem types.

Acknowledgements This material is based in part upon work supported by the National Science Foundation under Cooperative Agreement No. DBI-0939454. Any opinions, findings, and conclusions or recommendations expressed in this material are those of the author(s) and do not necessarily reflect the views of the National Science Foundation. This work was supported in part by Michigan State University through computational resources provided by the Institute for Cyber-Enabled Research.

References

1. Beadle, L., Johnson, C.: Semantically driven crossover in genetic programming. In: Wang, J. (ed.) Proceedings of the IEEE World Congress on Computational Intelligence, pp. 111–116. IEEE Computational Intelligence Society, IEEE Press, Hong Kong (2008). https://doi.org/10.1109/CEC.2008.4630784. http://results.ref.ac.uk/Submissions/Output/1423275
2. Burke, E.K., Gustafson, S., Kendall, G.: Diversity in genetic programming: an analysis of measures and correlation with fitness. IEEE Trans. Evol. Comput. **8**(1), 47–62 (2004). https://doi.org/10.1109/TEVC.2003.819263. http://www.cs.nott.ac.uk/~smg/research/publications/gustafson-ieee2004-preprint.pdf
3. Burks, A.R., Punch, W.F.: An efficient structural diversity technique for genetic programming. In: Silva, S., Esparcia-Alcazar, A.I., Lopez-Ibanez, M., Mostaghim, S., Timmis, J., Zarges, C., Correia, L., Soule, T., Giacobini, M., Urbanowicz, R., Akimoto, Y., Glasmachers, T., Fernandez de Vega, F., Hoover, A., Larranaga, P., Soto, M., Cotta, C., Pereira, F.B., Handl, J., Koutnik, J., Gaspar-Cunha, A., Trautmann, H., Mouret, J.B., Risi, S., Costa, E., Schuetze, O., Krawiec, K., Moraglio, A., Miller, J.F., Widera, P., Cagnoni, S., Merelo, J., Hart, E., Trujillo, L., Kessentini, M., Ochoa, G., Chicano, F., Doerr, C. (eds.) GECCO '15: Proceedings of the 2015 Annual Conference on Genetic and Evolutionary Computation, pp. 991–998. ACM, Madrid (2015). https://doi.org/10.1145/2739480.2754649. http://doi.acm.org/10.1145/2739480.2754649
4. De Jong, K.A.: Analysis of the behavior of a class of genetic adaptive systems. Ph.D. thesis, University of Michigan, Computer and Communication Sciences Department (1975)
5. de Jong, E., Watson, R., Pollack, J.: Reducing bloat and promoting diversity using multi-objective methods. In: Proceedings of the Genetic and Evolutionary Computation Conference, GECCO 2001, pp. 11–18. Morgan Kaufmann, New York (2001)

6. Gathercole, C., Ross, P.: The max problem for genetic programming - highlighting an adverse interaction between the crossover operator and a restriction on tree depth. Technical report, Department of Artificial Intelligence, University of Edinburgh, Edinburgh (1995). http://citeseer.ist.psu.edu/gathercole95max.html
7. Goldberg, D.E., Richardson, J.: Genetic algorithms with sharing for multimodal function optimization. In: Genetic Algorithms and Their Applications: Proceedings of the Second International Conference on Genetic Algorithms, pp. 41–49. Lawrence Erlbaum, Hillsdale (1987)
8. Helmuth, T., Spector, L., Matheson, J.: Solving uncompromising problems with lexicase selection. IEEE Trans. Evol. Comput. **19**(5), 630–643 (2015). https://doi.org/10.1109/TEVC.2014.2362729
9. Hornby, G.S.: Alps: the age-layered population structure for reducing the problem of premature convergence. In: Keijzer, M., Cattolico, M., Arnold, D., Babovic, V., Blum, C., Bosman, P., Butz, M.V., Coello Coello, C., Dasgupta, D., Ficici, S.G., Foster, J., Hernandez-Aguirre, A., Hornby, G., Lipson, H., McMinn, P., Moore, J., Raidl, G., Rothlauf, F., Ryan, C., Thierens, D. (eds.) GECCO 2006: Proceedings of the 8th Annual Conference on Genetic and Evolutionary Computation, vol. 1, pp. 815–822. ACM Press, Seattle (2006). https://doi.org/10.1145/1143997.1144142. http://www.cs.bham.ac.uk/~wbl/biblio/gecco2006/docs/p815.pdf
10. Hu, J., Seo, K., Li, S., Fan, Z., Rosenberg, R.C., Goodman, E.D.: Structure fitness sharing (SFS) for evolutionary design by genetic programming. In: Langdon, W.B., Cantú-Paz, E., Mathias, K., Roy, R., Davis, D., Poli, R., Balakrishnan, K., Honavar, V., Rudolph, G., Wegener, J., Bull, L., Potter, M.A., Schultz, A.C., Miller, J.F., Burke, E., Jonoska, N. (eds.) GECCO 2002: Proceedings of the Genetic and Evolutionary Computation Conference, pp. 780–787. Morgan Kaufmann Publishers, New York (2002). http://www.cs.bham.ac.uk/~wbl/biblio/gecco2002/GP195.pdf
11. Hu, J., Goodman, E., Seo, K., Fan, Z., Rosenberg, R.: The hierarchical fair competition framework for sustainable evolutionary algorithms. Evol. Comput. **13**(2), 241–277 (2005). https://doi.org/10.1162/1063656054088530
12. Jackson, D.: Phenotypic diversity in initial genetic programming populations. In: Esparcia-Alcazar, A.I., Ekart, A., Silva, S., Dignum, S., Uyar, A.S. (eds.) Proceedings of the 13th European Conference on Genetic Programming, EuroGP 2010. Lecture Notes in Computer Science, vol. 6021, pp. 98–109. Springer, Istanbul (2010)
13. Jackson, D.: Promoting phenotypic diversity in genetic programming. In: Schaefer, R., Cotta, C., Kolodziej, J., Rudolph, G. (eds.) PPSN 2010 11th International Conference on Parallel Problem Solving From Nature. Lecture Notes in Computer Science, vol. 6239, pp. 472–481. Springer, Krakow (2010)
14. Krawiec, K., O'Reilly, U.M.: Behavioral programming: a broader and more detailed take on semantic GP. In: Igel, C., Arnold, D.V., Gagne, C., Popovici, E., Auger, A., Bacardit, J., Brockhoff, D., Cagnoni, S., Deb, K., Doerr, B., Foster, J., Glasmachers, T., Hart, E., Heywood, M.I., Iba, H., Jacob, C., Jansen, T., Jin, Y., Kessentini, M., Knowles, J.D., Langdon, W.B., Larranaga, P., Luke, S., Luque, G., McCall, J.A.W., Montes de Oca, M.A., Motsinger-Reif, A., Ong, Y.S., Palmer, Y.S., Parsopoulos, K.E., Raidl, G., Risi, S., Ruhe, G., Schaul, T., Schmickl, T., Sendhoff, B., Stanley, K.O., Stuetzle, T., Thierens, D., Togelius, J., Witt, C., Zarges, C. (eds.) GECCO '14: Proceedings of the 2014 Conference on Genetic and Evolutionary Computation, pp. 935–942. ACM, Vancouver (2014). https://doi.org/10.1145/2576768.2598288. http://doi.acm.org/10.1145/2576768.2598288. Best paper
15. Langdon, W.B., Poli, R.: Foundations of Genetic Programming. Springer, Berlin (2002). https://doi.org/10.1007/978-3-662-04726-2. http://www.cs.ucl.ac.uk/staff/W.Langdon/FOGP/
16. Martinez, Y., Trujillo, L., Naredo, E., Legrand, P.: A comparison of fitness-case sampling methods for symbolic regression with genetic programming. In: Tantar, A.A., Tantar, E., Sun, J.Q., Zhang, W., Ding, Q., Schuetze, O., Emmerich, M., Legrand, P., Del Moral, P., Coello Coello, C.A. (eds.) EVOLVE - A Bridge Between Probability, Set Oriented Numerics, and Evolutionary Computation V. Advances in Intelligent Systems and Computing, vol. 288, pp. 201–212. Springer, Peking (2014)

17. McDermott, J., White, D.R., Luke, S., Manzoni, L., Castelli, M., Vanneschi, L., Jaskowski, W., Krawiec, K., Harper, R., De Jong, K., O'Reilly, U.M.: Genetic programming needs better benchmarks. In: Soule, T., Auger, A., Moore, J., Pelta, D., Solnon, C., Preuss, M., Dorin, A., Ong, Y.S., Blum, C., Silva, D.L., Neumann, F., Yu, T., Ekart, A., Browne, W., Kovacs, T., Wong, M.L., Pizzuti, C., Rowe, J., Friedrich, T., Squillero, G., Bredeche, N., Smith, S.L., Motsinger-Reif, A., Lozano, J., Pelikan, M., Meyer-Nienberg, S., Igel, C., Hornby, G., Doursat, R., Gustafson, S., Olague, G., Yoo, S., Clark, J., Ochoa, G., Pappa, G., Lobo, F., Tauritz, D., Branke, J., Deb, K. (eds.) GECCO '12: Proceedings of the Fourteenth International Conference on Genetic and Evolutionary Computation Conference, pp. 791–798. ACM, Philadelphia (2012). https://doi.org/10.1145/2330163.2330273
18. McKay, R.I.B.: Fitness sharing in genetic programming. In: Whitley, D., Goldberg, D., Cantu-Paz, E., Spector, L., Parmee, I., Beyer, H.G. (eds.) Proceedings of the Genetic and Evolutionary Computation Conference (GECCO-2000), pp. 435–442. Morgan Kaufmann, Las Vegas (2000). http://www.cs.bham.ac.uk/~wbl/biblio/gecco2000/GP256.pdf
19. McPhee, N.F., Hopper, N.J.: Analysis of genetic diversity through population history. In: Banzhaf, W., Daida, J., Eiben, A.E., Garzon, M.H., Honavar, V., Jakiela, M., Smith, R.E. (eds.) Proceedings of the Genetic and Evolutionary Computation Conference, vol. 2, pp. 1112–1120. Morgan Kaufmann, Orlando (1999). http://www.cs.bham.ac.uk/~wbl/biblio/gecco1999/GP-421.pdf
20. McPhee, N.F., Ohs, B., Hutchison, T.: Semantic building blocks in genetic programming. Working Paper Series, vol. 3(2), University of Minnesota Morris, Morris (2007)
21. Moraglio, A., Krawiec, K., Johnson, C.G.: Geometric semantic genetic programming. In: Coello Coello, C.A., Cutello, V., Deb, K., Forrest, S., Nicosia, G., Pavone, M. (eds.) Parallel Problem Solving from Nature, PPSN XII (Part 1). Lecture Notes in Computer Science, vol. 7491, pp. 21–31. Springer, Taormina (2012)
22. Nguyen, Q.U., Nguyen, X.H., O'Neill, M., Agapitos, A.: An investigation of fitness sharing with semantic and syntactic distance metrics. In: Moraglio, A., Silva, S., Krawiec, K., Machado, P., Cotta, C. (eds.) Proceedings of the 15th European Conference on Genetic Programming, EuroGP 2012. Lecture Notes in Computer Science, vol. 7244, pp. 109–120. Springer, Malaga (2012)
23. Nguyen, Q.U., Pham, T.A., Nguyen, X.H., McDermott, J.: Subtree semantic geometric crossover for genetic programming. Genet. Program Evolvable Mach. **17**(1), 25–53 (2016). https://doi.org/10.1007/s10710-015-9253-5
24. Poli, R., Langdon, W.B.: On the search properties of different crossover operators in genetic programming. In: Koza, J.R., Banzhaf, W., Chellapilla, K., Deb, K., Dorigo, M., Fogel, D.B., Garzon, M.H., Goldberg, D.E., Iba, H., Riolo, R. (eds.) Genetic Programming 1998: Proceedings of the Third Annual Conference, pp. 293–301. Morgan Kaufmann, University of Wisconsin, Madison (1998). http://www.cs.essex.ac.uk/staff/poli/papers/Poli-GP1998.pdf
25. Schmidt, M., Lipson, H.: Age-fitness Pareto optimization. In: Riolo, R., McConaghy, T., Vladislavleva, E. (eds.) Genetic Programming Theory and Practice VIII. Genetic and Evolutionary Computation, vol. 8, chap. 8, pp. 129–146. Springer, Ann Arbor (2010). http://www.springer.com/computer/ai/book/978-1-4419-7746-5
26. Spector, L., Clark, D.M., Lindsay, I., Barr, B., Klein, J.: Genetic programming for finite algebras. In: Keijzer, M., Antoniol, G., Congdon, C.B., Deb, K., Doerr, B., Hansen, N., Holmes, J.H., Hornby, G.S., Howard, D., Kennedy, J., Kumar, S., Lobo, F.G., Miller, J.F., Moore, J., Neumann, F., Pelikan, M., Pollack, J., Sastry, K., Stanley, K., Stoica, A., Talbi, E.G., Wegener, I. (eds.) GECCO '08: Proceedings of the 10th Annual Conference on Genetic and Evolutionary Computation, pp. 1291–1298. ACM, Atlanta (2008). https://doi.org/10.1145/1389095.1389343. http://www.cs.bham.ac.uk/~wbl/biblio/gecco2008/docs/p1291.pdf
27. Uy, N.Q., Hoai, N.X., O'Neill, M., McKay, R.I., Galvan-Lopez, E.: Semantically-based crossover in genetic programming: application to real-valued symbolic regression. Genet. Program Evolvable Mach. **12**(2), 91–119 (2011). https://doi.org/10.1007/s10710-010-9121-2

Chapter 3
Investigating Multi-Population Competitive Coevolution for Anticipation of Tax Evasion

Erik Hemberg, Jacob Rosen, and Una-May O'Reilly

Abstract We investigate the application of a version of Genetic Programming with grammars, called Grammatical Evolution, and a multi-population competitive coevolutionary algorithm for anticipating tax evasion in the domain of U.S. Partnership tax regulations. A problem in tax auditing is that as soon as one evasion scheme is detected a new, slightly mutated, variant of that scheme appears. Multi-population competitive coevolutionary algorithms are disposed to explore adversarial problems, such as the arms-race between tax evader and auditor. In addition, we use Genetic Programming and grammars to represent and search the transactions of tax evaders and tax audit policies. Grammars are helpful for representing and biasing the search space. The feasibility of the method is studied with an example of adversarial coevolution in tax evasion. We study the dynamics and the solutions of the competing populations in this scenario, and note that we are able to replicate some of the expected behavior.

Keywords Genetic programming · Tax evasion · Behavioral data mining · Coevolutionary algorithms · Adversarial learning

3.1 Introduction

We present an application of Genetic Programming (GP) on tax evasion, a fundamental and pervasive societal problem. Our focus is on the domain of U.S. Partnership taxation, which had a $91 billion tax gap in 2012 [10]. Tax evasion can be seen as an adversarial arms-race, the attacker is the tax evader and the defender is the auditor. Tax auditors have historical examples of tax schemes to help auditing. On the evasion side, tax shelter promoters often adapt their strategies as existing schemes are uncovered and when changes are made to the current tax

E. Hemberg (✉) · J. Rosen · U.-M. O'Reilly
MIT CSAIL, Cambridge, MA, USA
e-mail: hembergerik@csail.mit.edu; jbrosen@mit.edu; unamay@csail.mit.edu

R. Riolo et al. (eds.), *Genetic Programming Theory and Practice XIV*, Genetic and Evolutionary Computation,
https://doi.org/10.1007/978-3-319-97088-2_3

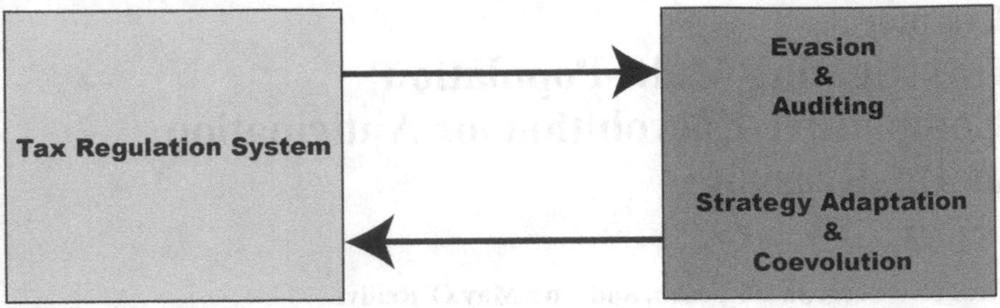

Fig. 3.1 STEALTH framework overview, tax regulation module and coevolutionary search module

regulations. One example is the so called BOSS tax shelter (Bond and Options Sales Strategies) that was widely promoted yet was ultimately disallowed. While changes were implemented to detect BOSS they were not able to detect the newly emerged variant "Son of BOSS" [29]. Previously, GP has been applied in law to study contracts as GP trees [5] and adversarial coevolutionary algorithms have been used to model arms-races, e.g. in cyber security [11].

There are some technical challenges to modeling U.S. partnership taxation: (A) the tax code's complexity, (B) the behaviors available to tax evaders and auditors, (C) the simultaneous co-adaptive behaviors of both auditors and tax evaders. First, we need to develop an abstraction of the relevant partnership tax law, which will allow us to compute both tax liability and likelihood of being audited. Second, we have to develop a model of taxpayer and auditor co-adaptive behavior. In the Simulation of Tax Evasion and Law Through Heuristics (STEALTH) framework (Fig. 3.1), we take knowledge from Coevolutionary algorithms [19] to model the dynamics of the coevolutionary adversarial relationship between tax evaders and auditors. Specifically the competitive coevolution with two populations, i.e. interactive test-based co-search. We chose a method with an explicit use of grammar to bias search and compress the description of the search space. Grammatical Evolution (GE) [18] is one such grammar based GP method. Grammars are appropriate start points for defining, constraining and adding domain knowledge to the search space of evader and auditor actions. In the experiments we analyze the adversarial search dynamics of applying STEALTH to a tax evasion scheme, called iBOB [9], designed to defer tax payments by using U.S. partnerships.

In Sect. 3.2 work on coevolution and GE are described. The STEALTH framework is described in Sect. 3.3. In Sect. 3.4 we describe our experiments using STEALTH. Finally, there are conclusions and future work in Sect. 3.5.

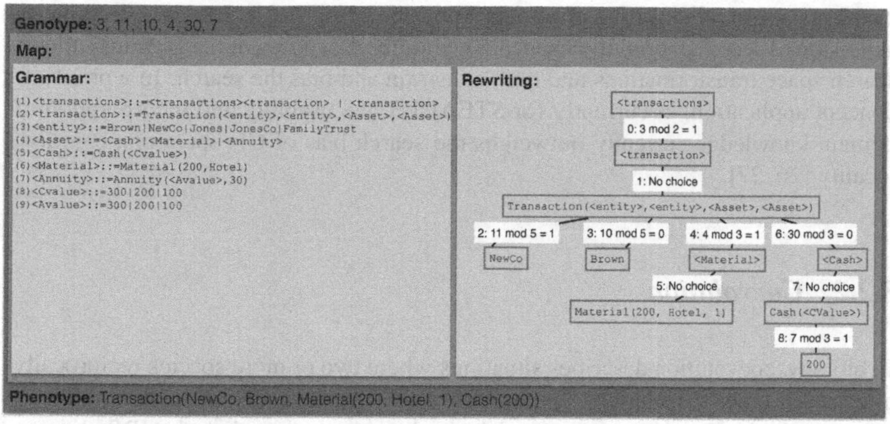

Fig. 3.2 Example of how GE rewrites a list of integers (genotype) into a list of transactions (phenotype) with a BNF grammar

3.2 Related Work

In the STEALTH framework for anticipating tax evasion we have used previous work in GE and coevolutionary algorithms. We describe the Genetic Programming approach with grammars called Grammatical Evolution in Sect. 3.2.1 and Coevolutionary algorithms in Sect. 3.2.2. Finally, the work on coevolution and GE are in Sect. 3.2.3.

3.2.1 Grammatical Evolution

Grammatical Evolution (GE) is a version of genetic programming with a variable-length integer representation and an indirect mapping using a grammar [18], which allows us to rewrite syntactically valid sentences (phenotypes), which in our case are transaction sequences and audit policies. As shown in Fig. 3.2, the grammar is composed of a single start symbol, terminal symbols and non-terminal symbols, as indicated by the boxes in the figure. The left hand box shows the grammar in BNF form and the right hand box shows the left-to-right rewriting of integers to sentences of only terminal symbols and the corresponding derivation tree. In the rewriting, the start symbol is at the top and terminal symbols are at the bottom of each branch of the derivation tree. Integers are first expanded at the root non-terminal, and then subsequent leaves, and the direction of the path is determined by taking the modulo of the current integer, at which point the next integer is selected. The process is complete when the sentence comprises only terminal symbols.

The use of a grammar has allowed GE to be applied to many different areas. For example, games, finance, design, hyper-heuristic for combinatorial optimization problems and parallel sorting programs on multi-cores [4, 6, 7, 22, 23]. GE provides a division of search and solution space. The indirect encoding of the grammar allows search space transformations and both constrain and bias the search. In a proof-of-concept application, as currently for STEALTH, the usability and incorporation of domain knowledge currently outweighs the search bias of GE operators, e.g., low locality [26, 27].

3.2.2 Coevolution

In biology, coevolution describes situations where two or more species reciprocally affect each other's evolution. The notion of adversarial coevolution from biology can be used for the circumstances of the auditors, e.g. each time the IRS changes the tax code the tax evaders react by finding new ambiguities. The auditor and the tax evaders are *coevolving* as interacting species. At its core, the overall dynamics of the system reflect a constantly transitioning series of complementary adjustments, with each predator/prey seeking to bring advantage to the predator/prey under adjustment.

Coevolutionary algorithms explore domains in which no single evaluation function is present or known. Instead, algorithms rely on the aggregation of outcomes from interactions among coevolving entities to make selection decisions. The lack of an explicit measurement, understanding the dynamics of coevolutionary algorithms, and determining the progress of a given algorithm present further challenges [19]. Usually, Evolutionary Algorithms begin with a fitness function, which for the purposes of this chapter is a function of the form $f : G \mapsto \mathbb{R}$ that assigns a real value to each possible solution (genotype) in G. Individual solutions are compared as $f(g_0)$ with $f(g_1)$ and the fitness ranking is always the same. In coevolution two individuals are compared based on their outcome from interaction with other individuals, thus the ranking of an individual solution can change over time.

Coevolution is appropriate for domains that have no intrinsic objective measure, also called *interactive* domains. There are two types of coevolution: **Compositional:** the quality of a solution to the problem involves an interaction among many components that together form a team, i.e. cooperating. **Test-based:** the quality of a potential solution is determined by its performance when interacting with some set of tests, i.e. competitive. For example, the interactive test for a tax evasion strategy are different auditor behaviors, and vice versa.

Coevolutionary algorithms and game theory are related, which leads to applications in games [19]. A distinction from game theory is that coevolutionary EAs can navigate large search spaces [21]. Some work has investigated solution concepts for test-case coevolution with a no-free-lunch framework [25]. In other applications, some of the problems with streaming data classification are encountered in coevolution [13], as well as comparing coevolution to boosting [14]. In addition, coevolution is also used for complexification of solutions [24]. Finally, there are also applications to simulations of behavior, e.g., Zero-Day exploit strategies in

Cyber Security [28] and the search for software bug fixes [15], where the program already exists. STEALTH addresses the EA "middle", between strategies and source code [21].

3.2.3 Coevolution and Grammatical Evolution

GE has been applied with coevolutionary algorithms. In the GE literature on dynamic environments [7] coevolution is characterized as "Markov", i.e. it has memory and proceeds from its current state. In coevolution the fitness of solutions are dependent on the search process, this dependency does not have to hold for dynamic environments. Some coevolutionary sources view interactive domains as static, with a different structure. Grammatical evolution by grammatical evolution (GEGE) [7] tries to simultaneously evolve the grammar and solutions with a hierarchy of grammars. The aim is to find "modules" that can be reused when the environment changes, i.e. in a dynamic environment. GEGE is a compact representation of a large grammar, with an increased search space, as well as strong coupling of the grammars [3].

Examples of applications of GE and coevolution are for reformulation of training as a two-population competition, that is learners versus training exemplars and use GE to represent Pareto-coevolutionary classifier and Multi-objective classifier [16, 17]. An Artificial Life model for evolving a predator–prey ecosystem of mathematical expressions with GE [2]. Coevolutionary algorithms with GE for financial trading, e.g., using multiple cooperative populations [1, 8]. As well as, spatial coevolution in age layered planes [12] for robots in robocode using competitive coevolution.

STEALTH requires the tax regulations to be encoded in software before the search for strategies can be performed. The tax law is defined by existing rules, in the IRC. The solutions in the separate populations in STEALTH are both interesting as well. Next we describes how to use GE and multi population interactive test based coevolution with U.S. partnership taxation.

3.3 Method

The STEALTH framework is composed of two modules: **Regulatory module:** the framework pertaining to U.S. partnership taxation. The legal logic to calculate tax liability and assign audit risk are here. **Coevolutionary module:** the algorithms which replicate the population dynamics with coevolutionary search. The values from the *regulatory module* are passed to the individual solutions from which they were derived. The modules are shown in Fig. 3.3.

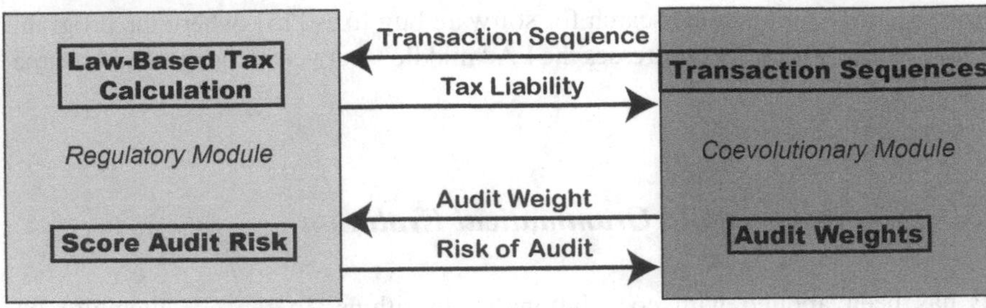

Fig. 3.3 STEALTH framework details

3.3.1 Tax Regulatory Module

The regulatory module provides the context in which the coevolutionary module, described in Sect. 3.3.2, is operating. Tax consequences of financial activity must be computed in order to analyze abusive tax behavior. Therefore, some aspects of Sub-chapter K of the IRC were hand coded into a formalism. This representation needs the ability to process an arbitrary set of financial behavior for tax liability.

Once we are able to calculate the tax liability resulting from a single transaction, we must determine what form a potentially abusive tax strategy takes, and how to represent a given auditing policy. The following sections describe our quantitative representations of tax evaders and auditors.

3.3.1.1 Tax Network and Transactions

We first define the environment, namely the initial conditions of a tax regulation unit. This amounts to a set of entities, each of which have a "portfolio" of assets, the entirety of which we refer to as the *tax network*. Transactions between entities are tax network state transitions. A transaction is considered a specific type of transition from one tax network state to another. A transaction is thus described as two entities and two assets that are being exchanged between the two entities. Finally we define a *transaction sequence* as a sequence of transactions.

3.3.1.2 Audit Score Sheets

We abstract a tax regulator for the tax regulatory model in order to determine the likelihood of conducting an audit, similar to how a tax regulated unit is represented as a transaction sequence. A regulator is a certain auditing policy, which is represented as a list of events *observable* within a transaction sequence with numerical weights associated with each type of observable event. When a particular observable event is applied to the regulatory module, the overall audit likelihood is

Table 3.1 Observable: the type of observable; Weight: the associated audit weight; Frequency: the number of times it occurs in a list of transactions

Observable	Weight	Frequency
Partnership interest sale (1)	w_1	f_1
No §754 election (2)	w_2	f_2
Substantial built-in loss (3)	w_3	f_3

incremented by its associated weight. The list of observable events with weights is referred to as the *audit score sheet*. The audit score sheet records the occurrence of each type of observable event, and any number of combinations of observable events. For example, consider the following passage from the Internal Revenue Code §743(*a*).

> The basis of partnership property shall not be adjusted as the result of (1) a transfer of an interest in a partnership by sale or exchange or on the death of a partner unless (2) the election provided by §754 (relating to optional adjustment to a basis of partnership property) is in effect with respect to such partnership of (3) unless the partnership has a substantial built-in loss immediately after such transfer.

Each number with parentheses signifies an *observable* event. Namely, (1) the sale of a partnership interest in exchange for a *taxable* asset; (2) the partnership whose shares are being transferred has not made a §754 election; (3) the seller's basis in respect to the non-cash assets owned by the partnership exceeds their FMV by more than $250,000. An audit score sheet that encapsulated only the three observable events listed in the passage would look as given in Table 3.1.

The formulation of the audit score sheet requires determining observables from the tax regulations. In contrast the calculation of the audit score sheet from is currently straightforward. We can write the audit score, s corresponding to the audit score sheet and network of transactions as $s = \sum_{i=0}^{n} \alpha_i f_i$ where $\sum_{i=0}^{n} \alpha_i = 1$.

3.3.2 Coevolutionary Module

The module which directs the appropriate adversarial dynamics has two subtasks. (A) assign *fitness* to both transaction sequences and audit score sheets, based on the measures of taxable income and audit score; (B) co-adapt the two competing populations—of solutions and tests—in terms of the fitness score, by searching over its behavioral space.

Our adversarial multi-population coevolutionary algorithm in STEALTH is in the interactive test domain [19]. The interaction is between populations of tax evasion schemes and audit policies. The tests are respectively audit observables and tax evasion scheme. A potential solution is therefore a tax evasion scheme (in the first case) or audit observable weights (in the second). The problem is to find the least risky evasion scheme or most likely audit observables. The domain is adversarial competition, the fitness of the tax evader is the opposite of the auditor. The problem is of the dual nature, i.e. the solutions for both sides are interesting.

3.3.2.1 Adversarial Population Representation

Individual solutions, i.e. transaction sequences and audit score sheets, are both evaluated in the tax regulatory system in separate populations. Therefore we must express and explore the spaces of all possible transaction sequences and audit weights.

We augment GE in a manner similar to GEGE to efficiently break up problem and gradually increase the search space of entities in STEALTH. Transaction sequences are generated by first removing some integers from the vector, which is a method that we use to generate initial tax network configurations. For example, if we would like to determine some number of additional partnerships in the initial configuration between 0 and k, we remove the first integer and take its modulo in respect to k, then process the rest of the vector through the grammar. Finally, mapping an integer sequence to an audit score sheet for an audit score sheet of length m we take an integer vector of length m and divide each integer in the vector by the sum of all of the elements. This creates m positive real numbers that sum to one.

3.3.2.2 Coevolutionary Tests: Objective Functions

We take into account both tax liability and likelihood of being detected by various auditing policies. Similarly, we search for an effective auditing policy. Neither task is trivial, nor can they be generalized to encompass every use-case. We apply a heuristic for determining effectiveness in a specific scenario to help formulate a good objective function. These heuristics are means to formulate objective functions for both a tax-minimizing strategy and audit weights, given a transaction sequence, initial tax network and audit score sheet.

An *objective function* is some mapping between the numerical traits associated with a transaction sequence or audit score sheet, and some measure of desirability. Section 3.3.1 describes how to calculate taxable income for all of the entities in the simulation, and an audit score. Given these two numerical constructs, we formulate objective functions for both transaction sequences, h_e, and audit score sheets, h_s, both are defined as a mapping from measure of taxable income and audit likelihood, to a real valued scalar.

An effective transaction sequence, from the perspective of a taxpayer, results in a low level of taxable income, with a low likelihood of being audited. A highly effective transaction sequence would be in the lower left corner, incurring relatively low levels of tax liability and little likelihood. A transaction sequence that produces low levels of tax liability but a *high* likelihood of being audited would be undesirable. We evaluate transaction sequences that all accomplish relatively similar economic goals. Thus any lower variations in taxable income can be indicative of, at the very least, tax implications that were never intended by policy-makers.

Auditing policies face a different heuristic for calculating effectiveness. Auditing policies must take into account the amount of resources that it takes to audit. An effective auditing policy avoid false positives and negatives and applies a low audit likelihood to transactions sequences that generated relatively normal levels of taxable income and a high likelihood to similar, low taxable income transaction sequences. Ineffective auditing policies are the exact opposite.

3.3.2.3 Adaptation—Coevolutionary Genetic Algorithm

The next task is to specify a means by which a large and highly non-linear space of transaction sequence-audit score sheet pairs can be co-adapted. The objectives establish a notion of effectiveness, the evolutionary algorithm determines (A) which transaction sequences can minimize tax liability while circumventing an audit and (B) which auditing policies assign high audit likelihood to relatively low tax liability schemes while ignoring non-suspicious behavior. The aim is to anticipate new forms of potentially abusive tax behavior as well as desirable, or likely, regulator response to it.

Upon establishing these mappings, a search can be performed on the space. Because of the predator–prey relationship between non-compliance schemes and auditing policy, we chose to use a *co-evolutionary algorithm*. Specifically, we evolve a test-based interaction problem. There are two competing populations of solutions that evolve in parallel and the fitness of a solution is subjective, i.e. the fitness depends on the test that the solution interacted with. Each individual in the two populations are evaluated against a subset of the opposing population. Our coevolutionary algorithm:

1. *initializes* both populations;
2. *evaluates* each individual against a subset of the other population members, to determine its objective score;
3. *selects* the best-scoring individuals in each population;
4. creates new populations by *crossover*, combining the selected individuals;
5. introduces slight *mutation* into the new population;
6. *repeats* steps 2–5 over some number of *generations*, until a halting condition occurs.

Specifically, every generation, each individual in the transaction sequence population selects a random subset of the audit score sheet of the population to evaluate against. After all sequences are evaluated, the process is repeated with the opposite population: each audit score sheet chooses a random subset of the transaction sequence population to evaluate. See [20] for more details.

3.4 Experiments

Ideally, we would like to be able to show that, with the proper specifications, dynamics between dominant tax strategies and dominant auditing policies can be replicated in a computational setting. That is, we see audit score sheets changing to

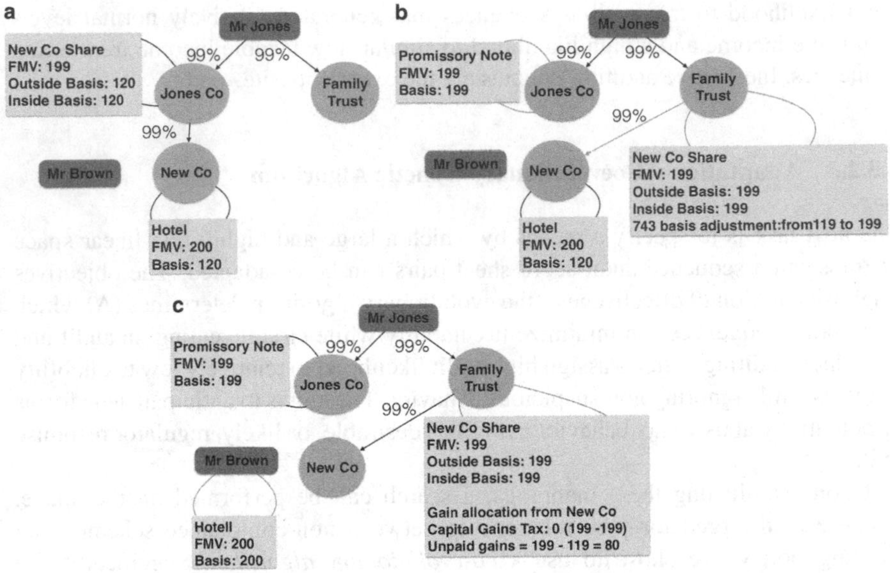

Fig. 3.4 The steps in the iBOB tax evasion scheme. The basis of an asset is artificially stepped up and tax is avoided by using "pass-through" entities. (**a**) iBob step 0. (**b**) iBob step 1. (**c**) iBob step 2

assign high audit likelihood to certain transaction sequence behavior that produces relatively low taxable income. Then in turn, we see the population of transaction sequences changing to favor transaction sequences that continue to produce low levels of taxable income, but using techniques that are not deemed suspicious by the dominant audit score sheets in the opposing population. We demonstrate population dynamics in STEALTH using a particular known tax evasion scheme called Installment Bogus Optional Basis (iBOB).

3.4.1 iBOB Description

In iBOB, a taxpayer arranges a network of transactions designed to reduce his tax liability upon the eventual sale of an asset owned by one of his subsidiaries [9]. He does this by stepping up the basis of this asset according to the rules set forth in §755 of the IRC. In this way, he manages to eliminate taxable gain while ostensibly remaining within the bounds of the tax law [29].

The sequence of transactions, shown graphically in Fig. 3.4, for the iBOB scheme are enumerated:

0. In the initial ownership network Mr. Jones is a 99% partner in JonesCo and FamilyTrust, whereas JonesCo is itself a 99% partner in another partnership,

NewCo. NewCo owns a hotel with a current fair market value (FMV) of \$200. If NewCo decides to sell the hotel at time step 1, Mr. Jones will incur a tax from this sale. The tax that Mr. Jones owes is the difference between the FMV at which the hotel was sold and his share of inside basis in this hotel, i.e. \$199 − \$119 = \$80. Mr. Jones can evade this tax by artificially stepping up the inside basis of the hotel to \$199.

1. In the first transaction, we see that FamilyTrust, which Mr. Jones controls, decides to buy JonesCo's partnership share in NewCo for a promissory note with a current value of \$199. Of course, FamilyTrust has no intention of paying off this note, as any such payments entail a tax burden upon NewCo. Having already made a 754 election, FamilyTrust steps up its inside basis in the hotel to \$199.
2. When NewCo sells the Hotel to Mr.Brown for \$200, Mr. Jones does not incur any tax, as the difference between the current market value and his share of inside basis in the hotel is now zero.

In our STEALTH implementation we note in addition to iBOB transaction (item 1) two additional patterns of transaction activity that can result in zero immediate tax liability for Mr. Jones. The first of these involves the transfer of a partnership interest between two "linked" entities in the same enterprise structure, usually resulting in a basis adjustment due to an earlier §754 election. By "linked" we mean a transaction in which the two parties are connected by an ownership relationship. In the iBOB context, these include "singly linked" transactions, such as those that may occur between Mr. Jones and JonesCo (or JonesCo and NewCo), and "doubly linked" transactions, as may occur between Mr. Jones and NewCo. These types of transactions result in zero immediate tax liability for all parties, but would almost certainly be audited. The second such transaction involves the use of Annuities such as promissory notes that are taxed only at the time of payment. As with "linked" transactions, defaulting on Annuity payments is nominally legal and results in zero tax liability but can be very suspicious for auditors.

3.4.2 Setup

We ran 100 independent iterations of the coevolutionary GA for 100 generations each with tax scheme and audit score populations of size 100. The parameters that govern the GA simulation are displayed in Table 3.2.

We are doing an initial exploration of the problem and the choice of parameters and operators are a first attempt. Each population has the same operator and parameter settings. In the initialization, integers are randomly chosen. Individuals are selected from the population using tournament selection. In the crossover operation two individuals are combined into two new individuals by randomly picking a single point and swapping after the point. The grammar used to map the integers in an individual is shown in Fig. 3.2 and the initial configuration is always the same. The mutation operation of an individual chooses a new random integer

Table 3.2 Parameters for STEALTH iBOB experiments

Parameter	Value	Description
Mutation rate	0.1	Probability of integer change in individual
Crossover rate	0.7	Probability of combining two individual integer strings
Tournament size	2	Number of competitors when selecting individuals
Number chosen	0.5	Fraction of other population each individual is evaluated
Population size	100	Number of individuals in each population
Generations	100	Number of times populations are evaluated

at a random position. The fitness of an individual is the average over a number of randomly chosen individuals from the other population.

The objective functions used for these experiments were single objective and opposites of one another. Given that Jones's taxable income was the only one of interest, $\zeta = 1$, and we can define his taxable income as just ℓ. Thus the objective function for a transaction sequence is $h_e(\ell, s) = -\ell(1 - s)$. Conversely, the objective function for audit score sheets are $h_a(\ell, s) = -\ell(1 - s)$. A primary assumption underlying our model is that tax schemes and audit scores sheets are engaged in a perpetual co-evolutionary process in which no global attractor exists. To generate sustaining oscillations, we constrain the resources of the auditor. We restrain the audit score sheets to assign the lowest audit point a value of zero, so that there will always be at least one scheme that is not detectable by the auditor. For the experiments considered the audit observables are `Material for Annuity`, `Single Linked`, `Double Linked`, and `iBOB`. The value of each audit point can be thought of as the relative importance of the associated transaction to the auditor.

3.4.3 Coevolution of Auditors & Evaders in iBOB

For our analysis of population dynamics we pick one run from our experiments. Fitnesses from various subpopulations of the transaction sequence population are shown in Fig. 3.5. A sharp increase in the fitness of the "best" transaction sequence indicated the discovery of a new way to minimize fitness. As soon as that occurs, the fitness of the best 10% of transaction sequences increases to the maximum, shortly followed by the fitness of all of the sequences in the population. The combined decline amongst all subpopulation fitnesses indicate the evolution of an audit score (not shown) sheet that increases the audit likelihood of a transaction sequence exhibiting the previously discovered scheme.

Thus the dynamics we set out to replicate with our model were displayed. Successful transaction sequences are those that generate low levels of taxable income for Jones, as well as exhibiting behavior that is not adequately represented in the audit score sheet population. Soon enough, the objective functions of the

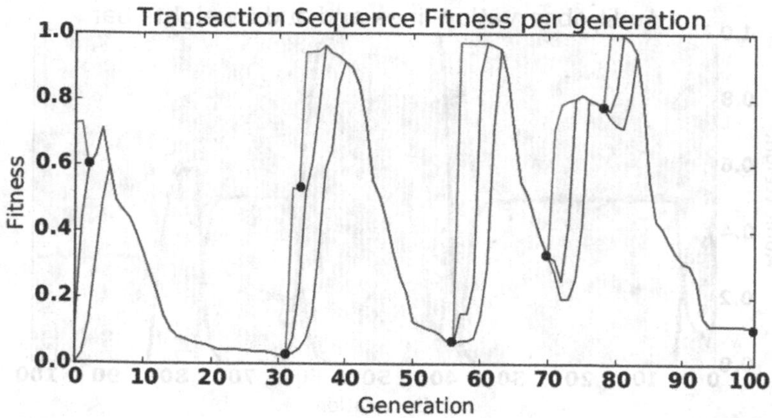

Fig. 3.5 Fitness of best transaction sequences (light grey), mean of top ten sequences (medium grey) and mean of population (dark grey). The dots signify points at which a novel tax-minimizing strategy is evolved

auditing policies begin to associate that behavior with low taxable income relative to other transaction sequences that accomplish the same economic purpose, and assign an audit weight to that behavior. The effectiveness of that tax strategy then decreases until a new tax-minimizing strategy is found which, once again evades all (or most) existing auditing policies. That strategy then rapidly spreads amongst the transaction sequence population, and the process continues.

Figure 3.6 below shows a nuanced picture of the audit score sheet population's response to the general trend in the transaction sequence population. The shaded background shows the audit weight distribution of the most fit audit score sheet in the population. Conversely, the shaded lines show the proportion of the transaction sequence population that uses the scheme of the corresponding shade. Thus we can see how the proportion of certain tax schemes follow the existence of the highest fitness audit score sheet.

We observe that an audit score sheet capable of sufficiently auditing a certain type of tax scheme can co-exist with that scheme for some time until the frequency of that tax strategy starts to decline. This demonstrates (a) the successful audit score sheet taking time to propagate amongst its population and (b) the notion of the fitness landscape of the transaction sequences. That is, audit score sheets have a fitness landscape that allows successful auditing policies to disseminate slowly. Conversely, dominant tax-minimization strategies have a more stochastic discovery process, but successful schemes propagate rapidly once found.

Table 3.3 show how the best transaction sequences change over generations for a run (the row number is shown in parentheses). The initial best transaction sequence is not a very large reduction in fitness (1), due to the random initialization of the populations. After two generations an improved sequence is found (2), although both a `Material for Annuity` and `Double Link` are observed in it. At generation seven (3) the best transaction sequence can only be audited with

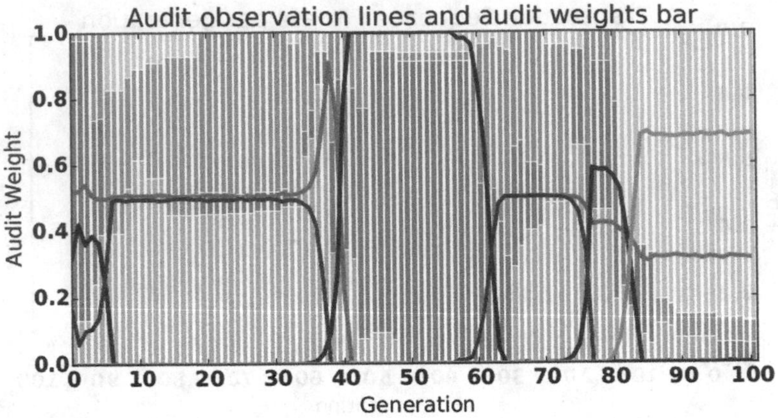

Fig. 3.6 Audit weights of best audit score sheet and proportions of various transaction sequence scheme types in population

`Material for Annuity`. By generation 31 (4) the best transaction sequence can be observed by a `Single Link` when Brown buys NewCo from JonesCo and then with the final transaction Brown buys the hotel from NewCo, which is observed as a single linked transaction. At generation 33 the best transaction (5) is possible to observe with a `Material for Annuity`. By generation 70 (6) the best transaction sequence is the same as at generation 31 (4) and can be observed by a `Single Link` when Brown buys the hotel from NewCo, the partnership he previously bough. Finally at generation 78 (7) the audit observable `iBOB` can capture the best transaction. We observe how all the audit observables were evaded due to the resource constraints on the auditor. It was also possible to note properties describe for co-evolutionary algorithms, namely how the evasion scheme (4) cycled in the population.

There are calibrations that can improve the fidelity of the experiments. For example, while transaction sequences are clearly more responsive to a successful individual in their population than audit score sheets, the time scale gives too much credit to the propagation of audit score sheets. For example, Fig. 3.5 shows that a successful tax strategy enjoys only about 5–10 generations until an auditing policy evolves and propagates that reduces its effectiveness. Transaction sequences take about the same amount of generations to figure out a new dominant tax strategy, the only tangible difference is the speed at which it propagates through the population. Thus, there must be further calibration in the model to reflect the differences in time scale.

Table 3.3 Best transactions sequences, the generation it entered the population and the audit observables required to detect it

Gen	Transactions	Audit observables	Row
< 2	Transaction(Brown, JonesCo, Annuity(200, 30), PartnershipAsset(99, NewCo)); Transaction(NewCo, Jones, Material(200, Hotel, 1), PartnershipAsset(99, JonesCo)	None (Not a good sequence)	(1)
< 7	Transaction(NewCo, Jones, Material(200, Hotel, 1), Annuity(300, 30))	`Material for annuity`, `Double linked`	(2)
< 31	Transaction(NewCo, FamilyTrust, Material(200, Hotel, 1), Annuity(300, 30))	`Material for annuity`	(3)
< 33	Transaction(Brown, JonesCo, Annuity(200, 30), PartnershipAsset(99, NewCo))	`Single linked`; (Brown buys hotel from himself)	(4)
< 70	Transaction(Brown, JonesCo, Annuity(200, 30), PartnershipAsset.(99, NewCo)); Transaction(NewCo, Jones, Material(200, Hotel, 1), Annuity(300,30))	`Material for annuity`	(5)
< 78	Transaction(Brown, JonesCo, Annuity(200, 30), PartnershipAsset(99, NewCo))	`Single linked` (same as at (4))	(6)
< 100	Transaction(FamilyTrust, JonesCo, Annuity(200, 30), PartnershipAsset(99, NewCo))	`iBOB`	(7)

Invalid transactions are cleared from the solutions for readability

3.5 Conclusions and Future Work

We presented how Genetic Programming, grammars and coevolutionary algorithms could be used for anticipating tax avoidance. The coevolutionary relationship between tax evaders and auditors was replicated with a coevolutionary algorithm, using U.S. partnership taxation as an initial example. We proceeded with (A) representing the rule system in order to calculate benefit that the advisor can offer to their client, (B) simulating interactions between the advisor's strategy and the relevant regulatory authority, and (C) optimizing for behavior on both ends of the relationship to investigate potential areas of exploration. Our experiments showed that the co-evolutionary relationship can be replicated, given the proper specifications. Some further parameter calibrations are required in order to capture certain time scale effects, but the qualitative dynamics are present. Transaction sequences can be shown to respond to both tax minimizing behavior and risk of being audited. Similarly, auditing policies respond to and isolate behavior which generates lower tax liability.

For future work we will expand our representation of the U.S. partnership tax code, e.g. non-recourse liabilities and depreciation deduction schedules. Another key aspect of validation is to gain access to actual auditing data. This is a non-trivial process that requires security clearance. Additional future work is to analyze the coevolutionary algorithm and dynamics (i.e. how the solutions in the populations are coevolving during the co-evolutionary search), the operators used for the search, pareto-archives, cycling of solutions and multi-objective fitness.

References

1. Adamu, K., Phelps, S.: Coevolutionary grammatical evolution for building trading algorithms. In: Electrical Engineering and Applied Computing, pp. 311–322. Springer, Dordrecht (2011)
2. Alfonseca, M., Gil, S.: Evolving a predator–prey ecosystem of mathematical expressions with grammatical evolution. Complexity **20**(3), 66–83 (2015)
3. Azad, R.M.A., Ryan, C.: An examination of simultaneous evolution of grammars and solutions. In: Genetic Programming Theory and Practice III, pp. 141–158. Springer, New York (2006)
4. Byrne, J., Cardiff, P., Brabazon, A., et al.: Evolving parametric aircraft models for design exploration and optimisation. Neurocomputing **142**, 39–47 (2014)
5. Chandler, S.J.: A 'genetically modified' liability insurance contract. University of Houston Law Center No. 2007-W-01 (2007)
6. Chennupati, G., Azad, R.M.A., Ryan, C.: Automatic evolution of parallel sorting programs on multi-cores. In: Applications of Evolutionary Computation, pp. 706–717. Springer, Cham (2015)
7. Dempsey, I., O'Neill, M., Brabazon, A.: Foundations in Grammatical Evolution for Dynamic Environments, vol. 194. Springer, Berlin (2009)
8. Gabrielsson, P., Johansson, U., Konig, R.: Co-evolving online high-frequency trading strategies using grammatical evolution. In: IEEE Conference on Computational Intelligence for Financial Engineering & Economics (CIFEr), 2104, pp. 473–480. IEEE, Piscataway (2014)
9. GAO: Gao-14-453 (2013). http://www.youtube.com/watch?v=O8VDUStvxMY
10. GAO (2014). http://www.gao.gov/assets/670/663185.pdf
11. Haddadi, F., Nur Zincir-Heywood, A.: Botnet detection system analysis on the effect of botnet evolution and feature representation. In: Proceedings of the Companion Publication of the 2015 on Genetic and Evolutionary Computation Conference, pp. 893–900. ACM, New York (2015)
12. Harper, R.: Evolving robocode tanks for evo robocode. Genet. Program Evolvable Mach. **15**(4), 403–431 (2014)
13. Heywood, M.I.: Evolutionary model building under streaming data for classification tasks: opportunities and challenges. Genet. Program Evolvable Mach. **16**(3), 283–326 (2015)
14. Iba, H.: Bagging, boosting, and bloating in genetic programming. In: Proceedings of the Genetic and Evolutionary Computation Conference. Morgan Kaufmann, Orlando (1999)
15. Le Goues, C., Nguyen-Tuong, A., Chen, H., Davidson, J.W., Forrest, S., Hiser, J.D., Knight, J.C., Van Gundy, M.: Moving target defenses in the helix self-regenerative architecture. In: Moving Target Defense II, pp. 117–149. Springer, Cham (2013)
16. McIntyre, A.R., Heywood, M.I.: Multi-objective competitive coevolution for efficient GP classifier problem decomposition. In: IEEE International Conference on Systems, Man and Cybernetics, 2007, ISIC, pp. 1930–1937. IEEE, Piscataway (2007)

17. McIntyre, A.R., Heywood, M.I.: Cooperative problem decomposition in Pareto competitive classifier models of coevolution. In: Genetic Programming, pp. 289–300. Springer, Berlin (2008)
18. O'Neill, M., Ryan, C.: Grammatical Evolution: Evolutionary Automatic Programming in an Arbitrary Language, vol. 4. Springer, New York (2003)
19. Popovici, E., Bucci, A., Paul Wiegand, R., De Jong, E.D.: Coevolutionary principles. In: Handbook of Natural Computing, pp. 987–1033. Springer, Berlin (2012)
20. Rosen, J.: Computer Aided Tax Avoidance Policy Analysis. Massachusetts Institute of Technology, Cambridge (2015)
21. Rush, G., Tauritz, D.R., Kent, A.D.: Coevolutionary agent-based network defense lightweight event system (candles). In: Proceedings of the Companion Publication of the 2015 on Genetic and Evolutionary Computation Conference, pp. 859–866. ACM, New York (2015)
22. Sabar, N.R., Ayob, M., Kendall, G., Qu, R.: Grammatical evolution hyper-heuristic for combinatorial optimization problems. IEEE Trans. Evol. Comput. **17**(6), 840–861 (2013)
23. Shaker, N., Nicolau, M., Yannakakis, G.N., Togelius, J., Neill, M.O.: Evolving levels for super mario bros using grammatical evolution. In: IEEE Conference on Computational Intelligence and Games (CIG), 2012, pp. 304–311. IEEE, Piscataway (2012)
24. Stanley, K.O., Miikkulainen, R.: Competitive coevolution through evolutionary complexification. J. Artif. Intell. Res. **21**, 63–100 (2004)
25. Tauritz, D.R. et al.: A no-free-lunch framework for coevolution. In: Proceedings of the 10th Annual Conference on Genetic and Evolutionary Computation, pp. 371–378. ACM, New York (2008)
26. Thorhauer, A., Rothlauf, F.: On the locality of standard search operators in grammatical evolution. In: Parallel Problem Solving from Nature–PPSN XIII, pp. 465–475. Springer, Cham (2014)
27. Whigham, P.A., Dick, G., Maclaurin, J., Owen, C.A.: Examining the best of both worlds of grammatical evolution. In: Proceedings of the 2015 on Genetic and Evolutionary Computation Conference, pp. 1111–1118. ACM, New York (2015)
28. Winterrose, M.L., Carter, K.M.: Strategic evolution of adversaries against temporal platform diversity active cyber defenses. In: Proceedings of the 2014 Symposium on Agent Directed Simulation, p. 9. Society for Computer Simulation International, San Diego (2014)
29. Wright, D. Jr.: Financial alchemy: How tax shelter promoters use financial products to bedevil the IRS (and how the IRS helps them). Ariz. State Law J. **45**, 611–675 (2013)

Chapter 4
Evolving Artificial General Intelligence for Video Game Controllers

Itay Azaria, Achiya Elyasaf, and Moshe Sipper

Abstract The General Video Game Playing Competition (GVGAI) defines a challenge of creating controllers for general video game playing, a testbed—as it were—for examining the issue of *artificial general intelligence*. We develop herein a game controller that mimics human learning behavior, focusing on the ability to generalize from experience and diminish learning time as *new* games present themselves. We use genetic programming to evolve hyper-heuristic-based general players. Our results show the effectiveness of evolution in meeting the generality challenge.

Keywords Genetic programming · Hyper-heuristics · Video games · GVG-AI competition

4.1 Introduction

Imagine playing a video game for the first time. You probably spend some time developing an understanding of the effects of your actions, identifying the various hostile or friendly non-player characters (NPCs), figuring out which items to avoid and which to collect, and so forth. In other words, you familiarize yourself with the goals of the game. A game can have many goals, such as collecting artifacts, killing NPCs, reaching a certain place in the game world, etc. There may be more complicated goals that are a combination of others, for example, collecting a key and then reaching the exit portal. Some of these goals lead to victory, some lead to award points, and some lead to both. Ultimately, we want to complete as many of these goals as possible before attempting the "end" goal that leads to victory and finishing the game, thus maximizing our score.

I. Azaria (✉) · A. Elyasaf · M. Sipper
Department of Computer Science, Ben-Gurion University, Beer-Sheva, Israel

R. Riolo et al. (eds.), *Genetic Programming Theory and Practice XIV*, Genetic and Evolutionary Computation,
https://doi.org/10.1007/978-3-319-97088-2_4

As you advance in the game you no longer have to focus on understanding it. Instead, you can focus on developing more advanced strategies or predicting the immediate future of the game, and estimating whether or not the strategies you have in mind are feasible.

When you move to other levels, games, or both, your previous experience helps to improve the learning curve. For example, you might not yet recognize the good and bad resources, but you already understand the concept of resources and non-player characters and the importance of keeping a certain distance from them. You will have to spend some time learning the specific new elements, or you might have to completely re-evaluate some insights—perhaps a wall is no longer an object you can't get through?—but you can rely on your previous experience to shorten the learning process.

Our goal in this work is to develop an artificial intelligence *game controller* (or *player*) that mimics this behavior, specifically focusing on the ability to generalize from experience and diminish learning time as new games present themselves.

The General Video Game Playing Competition [18] proposes the challenge of creating a controller for general video game playing, as a testbed for examining the issue of artificial general intelligence.

The GVGAI competition provides a framework that comprises a set of games, which differ in various aspects, including: winning conditions, scoring mechanism, sprite types, and available actions. The world the agent plays in is fully observable, and a forward model is provided. However, the games are stochastic and no information is provided regarding winning conditions or interactions between different elements in the world. It is up to the agent to either infer such information or otherwise search in the state space.

Our goal is *not* to win the GVGAI competition (even though our results turned out to be excellent), but to use the offered framework as a convenient, extant benchmark for our work. This decision stems partly from a technical reason (involving unsupported multi-threading, as elaborated in Sect. 4.3.4) that prevents us from competing online directly, and also from a desire to emphasize the *general* part of AGI, whereupon we wished to avoid specificity as much as possible.

The chapter is organized as follows: In the next section we examine previous and related work. In Sect. 4.3 we describe our method. Finally, we end with concluding remarks and future work in Sect. 4.4.

4.2 Previous Work

4.2.1 Automated Planning and MDP

Automated planning is a field of research in which generalized problem solvers (known as *planning systems* or *planners*) are constructed and tested across various benchmark puzzle domains.

A Markov decision process (MDP) is a model of an agent interacting synchronously with some given "world". The agent takes as input the state of the world and generates actions as output, which themselves affect the state of the world. While there is uncertainty in the MDP framework regarding the outcome of the agent's actions, the agent's current state is known. The description of MDP problems usually includes: (1) a transition function, assigning probabilities of reaching each state after performing a specific action in a given state; and (2) a reward function, describing the immediate reward for performing an action at a certain state [14].

The problem we face is similar in that the world can be represented as a set of states, and the agent must choose from a set of actions according to the current state. However, we have neither the transition nor the reward function. Moreover, since we are not allotted time for pre-computation, it is not possible to derive estimations of these functions.

4.2.2 Heuristic Search

Search algorithms for video controllers (as well as for other types of problems) are strongly based on the notion of approximating the distance of a given configuration (or *state*) to the *goal* (e.g., maximum score, catching the flag, etc.). Such approximations are found by means of a computationally efficient function, known as a *heuristic function*. By applying such a function to states reachable from the current one considered, it becomes possible to select more-promising alternatives earlier in the search process, possibly achieving better results (e.g., a higher score, or reaching the goal faster). The putative result is strongly tied to the quality of the heuristic function used: employing a perfect function means simply "strolling" onto the solution (i.e., no search de facto) and maximizing the solution score, while using a bad function could render the search less efficient than totally uninformed search, such as breadth-first search (BFS) or depth-first search (DFS).

4.2.3 Hyper-Heuristics

In the area of combinatorial optimization the term hyper-heuristics was first used by Cowling et al. [7] to describe heuristics to choose heuristics. This definition of hyper-heuristics was expanded later [4] to refer to an automated methodology for selecting or generating heuristics to solve hard computational search problems. In the process of hyper-heuristics learning, heuristics are used as building blocks. These heuristics can be of high level, usually complex and memory-consuming (e.g., landmarks and pattern databases), or low-level heuristics that are usually intuitive and straightforward to implement and compute.

Hyper-heuristics have been applied in many research fields, among them:

- Classical planning [11, 15, 21].
- Classical NP-Complete domains, e.g., 2D and 3D bin-packing [5, 6], personnel scheduling [3, 7].
- Classical AI domains and puzzles, e.g., the Rush Hour puzzle [12], the game of FreeCell [9], and the Tile Puzzle [1, 8].
- Mining RNA sequence-structure motifs [10].

The growing research interest in techniques for automating the design of heuristic search methods motivates the search for automatic systems for generating hyper-heuristics.

4.2.4 Real-Time Learning of Hyper-Heuristics

While research on learning hyper-heuristics is numerous, there is little work on real-time learning of hyper-heuristics. In many cases, converting offline algorithms to real-time ones is not trivial because real-time algorithms must handle the rapid change of the domain and the problems, while maintaining previous knowledge.

4.2.5 Solvers from GVGAI (Monte Carlo)

Many Monte Carlo Tree Search (MCTS) [2] submissions often outperformed other alternatives, with even the given, sample MCTS algorithm ranking third on the 2014 GVGAI competition.

Top performers included fast evolutionary MCTS and knowledge-based fast evolutionary MCTS [17], which embedded the algorithm roll-outs within evolution. The individuals of the evolutionary algorithm were weight vectors, used to bias the roll-outs of MCTS. Every roll-out performed during the search evaluated a single individual of the evolutionary algorithm, providing as fitness the reward calculated at the end of the roll-out.

Also of note is MCTS with influence map [16], the latter element being a numerical representation of influence on the game map, helping find a road to rewards over the horizon. The influence map essentially assigns a value to each object in the game world, representing if it is ‘good’ or ‘bad’. The value is then updated upon interaction with the object, eventually causing the player to focus on rewarding interactions.

4.3 Method

Recall our discussion on playing a video game for the first time. Our goal herein is to apply AI real-time learning techniques to develop controllers for video games that mimic the ability to generalize from previous experience, and reduce learning time as new levels present themselves.

Our approach comprises two principle components working in parallel as the game-playing agent encounters new levels:

1. Learning the heuristic function for evaluating the states' potential for achieving the different goals. The output of this component is a hyper-heuristic that receives a game state and returns a linear combination of different game goals, representing the potential for achieving these goals.
2. A game controller that uses generated hyper-heuristics in order to pick the most promising course of action. During its run the game controller passes encountered game states to the learning component. The latter learns the given states and incorporates the extracted knowledge to the evolved hyper-heuristic.

4.3.1 Heuristic Templates

Many games' goals can be generalized into the same principal goal, with minor variations. The most straightforward example is goals that relate to distances in the game, as in Pacman and Zelda, two seemingly different games [18]. In Pacman, the main goal is to clear the board, i.e., "eat" all the pills and power pills. Bonus points are given for "eating" ghosts while under the effect of a power pill. In Zelda, the goal is to collect the key, then exit the level. Here, killing enemies with the sword rewards the player with bonus points.

The goals of collecting pills/keys in Pacman/Zelda, respectively, are computationally identical—we need to minimize the distance between the player and the objects, represented as number of steps. Similarly, we have the goal of maintaining a certain distance from either ghosts/enemies in Pacman/Zelda, respectively. The mechanism of calculating the distance is identical, however, the conditions regarding when to minimize or maximize it are different: in Pacman we wish to minimize the distance if we are under the effect of a power pill, and maximize otherwise; in Zelda we want to minimize distance if we can use our sword in the appropriate direction, and maximize it otherwise, if we are short on time and need to exit the level.

Heuristic templates are our method of encoding knowledge that is relevant to most game domains, in a way that can be customized as the game runs. They can be viewed as parameterized heuristics, where the parameters are the current state and any number of additional integers representing: (a) a specific object in the game world; (b) a type of an object in the game world; or (c) a list (enum).

Table 4.1 Heuristic templates

Id	Name	Parameters	Description
1	Heuristic distance between object types	Integer type 1, Integer type 2	Parameterized by two types of objects. Values are taken from a set of types the controller encountered during the game. Calculates the geometric mean for each sprite type, and returns the Manhattan distance between the two means
2	Count nearby NPCs	Integer Manhattan distance	Counts the number of NPCs that are less than Manhattan distance away from the player
3	Score in K moves	Integer *k*	Game score if the controller does nothing for *k* moves. Useful for predicting how "safe" a position is
4	Distance from corner	enum *corner*	If there is a path between *corner* and the player, returns its length. Otherwise, returns the Manhattan distance between the player and the corner
5	Movable distance from immovable	Integer *immovable*	Calculates the A* distances between the selected immovable and all movable objects
6	Heuristic distances	Object reference, list objects	Calculates the Manhattan distances between the given reference object and the list of objects. The reference object/Objects list can be any of the ones exposed by the state

A complete list of our heuristic templates is given in Table 4.1. Additional non-parameterized heuristics are given in Table 4.2.

4.3.2 Hyper-Heuristics

Combining several heuristics to get a more accurate one is considered one of the most difficult problems in contemporary heuristics research [4, 20].

This task typically involves solving two major sub-problems:

1. How to combine heuristics by *arithmetic* means, e.g., by summing their values or taking the maximal value.

Table 4.2 Non-parameterized heuristics

Id	Name	Description
7	Facing NPC	Equals 1 if an NPC is directly in front of the player, −1 otherwise
8	Avatar resources count	The total number of resources the player possesses
9	Immovable count	The total number of immovable objects that are not walls, in the game world
10	NPC count	The total number of NPCs in the game world
11	Last action is 'use'	Equals 1 if the last action successfully performed by the player was 'use'. Can be used to reward applying that action in games where that is beneficial
12	Touching walls count	Counts the number of walls blocking the player (0–3)
13–18	A* distance	A list of the distances—measured in number of states—between the player and one of: NPCs, immovable objects, portals, resources, movable objects. The distance is calculated as follows: at the beginning of each level, a graph is generated from the world, with a node for each position the player can stand on. Nodes are connected if they are adjacent. Upon a request for a distance between two nodes, a path is calculated using weighted A* [19] and then cached for the duration of the level

2. Finding exact conditions (i.e., *logic* functions) regarding *when* to apply each heuristic, or combinations thereof—some heuristics may be more suitable than others when dealing with specific state.

In order to accomplish the first task, we first need to generate heuristics from our heuristic templates. To do that we differentiate between the different return values of the templates. If the template returns a real number, the generated heuristic is the returned value multiplied by a fixed weight, randomized during the heuristic generation. If the returned value is a list of real numbers, we first have to perform one of the following aggregating arithmetic: min, max, sum, multiplication, or division. Similarly, we do the same for basic heuristics.

Our learning algorithm solves the problem of combining heuristics by quick evaluation, thereby ruling out unpromising hyper-heuristics. Finding the exact condition under which to apply each hyper-heuristic is solved by design—new hyper-heuristics are evaluated specifically on game states the controller encountered recently, and often states the controller will encounter shortly.

4.3.3 *Learning Hyper-Heuristics Through Evolution*

4.3.3.1 Individuals

Our individuals are hyper-heuristics. An individual is composed of a list of heuristics derived from heuristic templates, basic heuristics, and basic observations. A hyper-heuristic is a linear combination of these heuristics.

An individual is represented as an *abstract syntax tree* (AST) [13] of a Java class, which is compiled during evaluation.

4.3.3.2 Fitness Function

As the real-time algorithm imposes fast hyper-heuristic evaluations, we expand the given initial state to the depth of X and set the individual fitness to be: $h(final\ state) - h(initial\ state)$ where h is the individual. In order to return values in a timely manner X was chosen to be 70.

At each depth we choose the next action in the following manner: We perform a 3-step-look-ahead by applying all possible actions to the depth of three. We choose the action that—on average—led to the highest heuristic value.

We used standard, Koza-style GP with the usual suspects: tournament selection with group size $k = 3$; subtree crossover; and constant and subtree-grow mutation operators.

4.3.4 *GVGAI*

As discussed in Sect. 4.1, the GVGAI competition proposes the challenge of creating controllers for a large range of stochastic real-time games, allowing the participants to train on a set of games, while testing them on a different, undisclosed set of games.

The competition imposes a single-thread limit, preventing us from participating because we need multi-threading for our online learning. Instead, we use the framework provided for the competition as a benchmark for our learning algorithm. We note that while the element of the undisclosed games is removed, we do not perform any sort of training on a subset of the games, or perform adjustments for specific games; instead we focus on real-time learning of the scenarios presented to the controller.

The GVGAI framework works by presenting the controller with a state that represents the fully observable world. The controller then has 40 ms to return a selected action. Along with the state a forward model is provided in the form of a search tree, where each node represents a state, and each edge an action. Being of stochastic nature, performing action A from state S may yield different results whenever we attempt it.

4.3.5 Game Controller

As discussed above, under the GVGAI framework our controller has 40 ms to return an action for a given state. We first handle communication with the learning component: We save the current state for future GP generations, and take the best hyper-heuristic learned so far for our current evaluation.

We then use a 3-step-lookahead search algorithm that is almost identical to the one used during fitness calculation: expand the search tree to depth 3 and calculate the highest heuristic score reachable for each possible move. However, unlike during fitness calculation, we repeat this step, taking the average heuristic score, thus avoiding moves that are not likely to constantly lead to desirable moves.

4.4 And the Winner is . . .

Though a comparison with the 2014 entries would be somewhat of an "apples and oranges" case, given our intentional deviation from the strict GVGAI framework, we still note—with some pride—that we would have ranked number three out of nineteen competitors.

For testing, we used the three game sets used in the GVGAI competition in CIG 2014—which are now fully available as training sets. GVGAI used a formula-one like scoring system, awarding scores for the best performing contenders for each game. However, since we're mostly interested in the generality of our algorithm, and not in individual game performance, we compare it using percent of games won.

In the CIG 2014 GVGAI competition our algorithm would have ranked third out of 19, with 36.3% of games won. If we increase the time the controller has before it must return an action from 40 to 80 ms, this percentage increases to 40.09% games won, leading us to believe our method is scalable, and would perform better with more resources.

Though the road ahead still has many paths to follow, we believe we have begun meeting our challenge: using genetic programming to evolve hyper-heuristic-based *general* players through real-time, online learning.

References

1. Arfaee, S.J., Zilles, S., Holte, R.C.: Bootstrap learning of heuristic functions. In: Proceedings of the 3rd International Symposium on Combinatorial Search (SoCS2010), pp. 52–59 (2010)
2. Browne, C.B., Powley, E., Whitehouse, D., Lucas, S.M., Cowling, P.I., Rohlfshagen, P., Tavener, S., Perez, D., Samothrakis, S., Colton, S.: A survey of monte carlo tree search methods. IEEE Trans. Comput. Intell. AI Games **4**(1), 1–43 (2012)

3. Burke, E.K., Kendall, G., Soubeiga, E.: A tabu-search hyperheuristic for timetabling and rostering. J. Heuristics **9**(6), 451–470 (2003). https://doi.org/10.1023/B:HEUR.0000012446.94732.b6
4. Burke, E.K., Hyde, M., Kendall, G., Ochoa, G., Ozcan, E., Woodward, J.R.: A classification of hyper-heuristic approaches. In: Gendreau, M., Potvin, J. (eds.) Handbook of Meta-Heuristics, 2nd edn., pp. 449–468. Springer, Boston (2010)
5. Burke, E.K., Hyde, M.R., Kendall, G., Woodward, J.: A genetic programming hyper-heuristic approach for evolving 2-D strip packing heuristics. IEEE Trans. Evol. Comput. **14**(6), 942–958 (2010). https://doi.org/10.1109/TEVC.2010.2041061
6. Burke, E.K., Hyde, M.R., Kendall, G., Woodward, J.: Automating the packing heuristic design process with genetic programming. Evol. Comput. **20**(1), 63–89 (2012). https://doi.org/10.1162/EVCO_a_00044
7. Cowling, P.I., Kendall, G., Soubeiga, E.: A hyperheuristic approach to scheduling a sales summit. In: Burke, E.K., Erben, W. (eds.) PATAT 2000: Practice and Theory of Automated Timetabling III. Lecture Notes in Computer Science, vol. 2079, pp. 176–190. Springer, Berlin (2000). https://doi.org/10.1007/3-540-44629-X_11
8. Elyasaf, A., Zaritsky, Y., Hauptman, A., Sipper, M.: Evolving solvers for FreeCell and the sliding-tile puzzle. In: Borrajo, D., Likhachev, M., López, C.L. (eds.) Proceedings of the Fourth Annual Symposium on Combinatorial Search, SoCS 2011, Castell de Cardona, Barcelona, Spain, July 15–16, 2011. AAAI Press, Palo Alto (2011). http://www.aaai.org/ocs/index.php/SOCS/SOCS11/paper/view/4018
9. Elyasaf, A., Hauptman, A., Sipper, M.: Evolutionary design of FreeCell solvers. IEEE Trans. Comput. Intell. AI Games **4**(4), 270–281 (2012). https://doi.org/10.1109/TCIAIG.2012.2210423. http://ieeexplore.ieee.org/xpls/abs_all.jsp?arnumber=6249736
10. Elyasaf, A., Vaks, P., Milo, N., Sipper, M., Ziv-Ukelson, M.: Learning heuristics for mining RNA sequence-structure motifs. In: Genetic Programming Theory and Practice XIII (GPTP 2015). Springer, Cham (2015)
11. Fawcett, C., Karpas, E., Helmert, M., Roger, G., Hoos, H.: Fd-autotune: domain-specific configuration using fast-downward. In: Proceedings of ICAPS-PAL 2011 (2011)
12. Hauptman, A., Elyasaf, A., Sipper, M., Karmon, A.: GP-rush: using genetic programming to evolve solvers for the Rush Hour puzzle. In: GECCO'09: Proceedings of 11th Annual Conference on Genetic and Evolutionary Computation Conference, pp. 955–962. ACM, New York (2009). https://doi.org/10.1145/1569901.1570032. http://dl.acm.org/citation.cfm?id=1570032
13. Jones, J.: Abstract syntax tree implementation idioms. In: Proceedings of the 10th Conference on Pattern Languages of Programs (plop2003), pp. 1–10 (2003)
14. Kaelbling, L.P., Littman, M.L., Cassandra, A.R.: Planning and acting in partially observable stochastic domains. Artif. Intell. **101**(1), 99–134 (1998)
15. Levine, J., Westerberg, H., Galea, M., Humphreys, D.: Evolutionary-based learning of generalised policies for AI planning domains. In: Rothlauf, F. (ed.) Proceedings of the 11th Annual Conference on Genetic and Evolutionary Computation (GECCO 2009), pp. 1195–1202. ACM, New York (2009)
16. Park, H., Kim, K.J.: MCTS with influence map for general video game playing. In: IEEE Conference on Computational Intelligence and Games (CIG), 2015, pp. 534–535. IEEE, Piscataway (2015)
17. Perez, D., Samothrakis, S., Lucas, S.: Knowledge-based fast evolutionary MCTS for general video game playing. In: IEEE Conference on Computational Intelligence and Games (CIG), 2014, pp. 1–8. IEEE, Piscataway (2014)
18. Perez, D., Samothrakis, S., Togelius, J., Schaul, T., Lucas, S., Couëtoux, A., Lee, J., Lim, C.U., Thompson, T.: The 2014 general video game playing competition. IEEE Trans. Comput. Intell. AI Games **8**, 229–243 (2015)

19. Pohl, I.: Heuristic search viewed as path finding in a graph. Artif. Intell. **1**(3), 193–204 (1970)
20. Samadi, M., Felner, A., Schaeffer, J.: Learning from multiple heuristics. In: Fox, D., Gomes, C.P. (eds.) Proceedings of the Twenty-Third AAAI Conference on Artificial Intelligence (AAAI 2008), pp. 357–362. AAAI Press, Palo Alto (2008)
21. Yoon, S.W., Fern, A., Givan, R.: Learning control knowledge for forward search planning. J. Mach. Learn. Res. **9**, 683–718 (2008). http://doi.acm.org/10.1145/1390681.1390705

Chapter 5
A Detailed Analysis of a PushGP Run

Nicholas Freitag McPhee, Mitchell D. Finzel, Maggie M. Casale, Thomas Helmuth, and Lee Spector

Abstract In evolutionary computation we potentially have the ability to save and analyze every detail in an run. This data is often thrown away, however, in favor of focusing on the final outcomes, typically captured and presented in the form of summary statistics and performance plots. Here we use graph database tools to store every parent–child relationship in a single genetic programming run, and examine the key ancestries in detail, tracing back from an solution to see how it was evolved over the course of 20 generations. To visualize this genetic programming run, the ancestry graph is extracted, running from the solution(s) in the final generation up to their ancestors in the initial random population. The key instructions in the solution are also identified, and a genetic ancestry graph is constructed, a subgraph of the ancestry graph containing only those individuals that contributed genetic information (or instructions) to the solution. These visualizations and our ability to trace these key instructions throughout the run allow us to identify general inheritance patterns and key evolutionary moments in this run.

Keywords Genetic programming · Visualization · Ancestry analysis · Push language · Evolutionary dynamics

---

N. F. McPhee (✉) · M. D. Finzel · M. M. Casale
University of Minnesota, Morris, Morris, MN, USA
e-mail: mcphee@morris.umn.edu; finze008@morris.umn.edu; casal033@morris.umn.edu

T. Helmuth
Computer Science, Washington and Lee University, Lexington, VA, USA
e-mail: helmutht@wlu.edu

L. Spector
Cognitive Science, Hampshire College, Amherst, MA, USA
e-mail: lspector@hampshire.edu

R. Riolo et al. (eds.), *Genetic Programming Theory and Practice XIV*, Genetic and Evolutionary Computation,
https://doi.org/10.1007/978-3-319-97088-2_5

5.1 Introduction

Previous work [1–5, 9, 10] has illustrated the value of ancestry graphs as a means of analyzing the dynamics of evolutionary computation runs. In [10], for example, we demonstrated the use of graph databases as a tool for collecting and analyzing ancestries in genetic programming runs, identifying several key moments and general patterns in runs using both lexicase and tournament selection.

In this paper we extend that work to provide a more detailed analysis of a single, complete run. We identify *every* ancestor of the evolved solutions, and then reduce that graph (which has 394 individuals) to a graph containing only the individuals that in fact contributed one of the key instructions to the final solutions (73 individuals). We then trace each of these key solution instructions back through the entire lineage, identifying where they were first introduced, and how they were transmitted through the genetic history. This reveals a number of interesting properties of this particular run including, for example, the fact that four of the nine key instructions were introduced via mutation and most crossover events led to changes that could have been brought about by mutation alone.

In Sect. 5.2 we review the key components of the system used to generate the run explored here (PushGP, Plush genomes, and the Replace Space With Newline test problem). We then describe and present both the full and genetic ancestry graphs in Sect. 5.3, before tracing the evolutionary history of all the key instructions in Sect. 5.4. Our discussion in Sect. 5.5 builds on the details of these traces and catalogues the kinds of events we see in this run, describing a few in greater detail. We then wrap up with some conclusions and ideas for future work in Sect. 5.6.

5.2 Languages, Configuration, Tools and Setup

The run presented here was generated using a Clojure implementation[1] of the PushGP[2] genetic programming system, which evolves programs in the Push programming language [12, 15]. Push programs use typed stacks to store and manipulate data, taking their arguments from stacks of the appropriate types and leave their results on the appropriate stacks.

Push gains much of its power as an evolutionary language from its ability to manipulate code, including the currently executing code, as a program runs. The running program is stored on the *exec* stack, allowing instructions to change code before it runs. Push programs are hierarchically structured into code blocks delimited by parentheses. Each code block is treated as a single unit when code manipulating instructions act on them.

[1] Clojush: https://github.com/lspector/Clojush.

[2] http://pushlanguage.org/.

Unlike previous versions of PushGP, Clojush has recently been changed to not evolve Push programs directly, but to act instead on a new linear genome representation [7]. Each *Plush (Linear **Push**) genome* consists of a linear sequence of instructions (including literals), and is translated into a Push program prior to execution. Each instruction may have one or more epigenetic markers attached that modify how the genome is translated into a Push program. For more details on the Plush genome representation and operators, see [7].

Most relevant to this study is the *close* epigenetic marker, which affects the hierarchical composition of programs. Since many Push instructions do not act on code blocks from the *exec* stack, it makes sense to limit the appearance of code blocks to follow only instructions that do make use of them. Each instruction that takes one or more argument from the *exec* stack automatically opens one or more code blocks. Then, the integer close marker attached to each instruction tells how many opened code blocks to close after that particular instruction. During translation from Plush genome to Push program, an open parenthesis is placed after each instruction that requires a code block, and a matching closing parenthesis is placed after a later instruction with a non-zero close marker. These code blocks can create hierarchically nested Push programs, allowing, for example, structures such as nested looping and subroutines containing conditional code.

This run used *lexicase selection* [6, 8, 11]. The details of lexicase selection aren't crucial here, but it is important to know that lexicase selection avoids aggregating test case performance (by, for example, computing a single total error as is common when using tournament selection), and instead maintains a vector of distinct errors for each test case. This allows an individual that performs well on a few test cases that the population is generally poor at to be selected, often multiple times, even if it performs very poorly on other test cases.

Our main crossover operator, *alternation*, similar to N-point crossover in genetic algorithms. Alternation traverses two parents in parallel while copying instructions from one parent or the other to the child. While traversing the parents, copying can jump from one parent to the other with probability specified by the *alternation rate* parameter. When alternating between parents, the index at which to continue copying may be offset backward or forward some small random amount.

We also use a *uniform mutation* operator that traverses a parent, replacing each instruction with some small probability. Similarly, a *uniform close mutation* operator can change the close epigenetic marker attached to an instruction by incrementing or decrementing it. Finally, we often apply an alternation operator followed by a uniform mutation of the result, inspired by the ULTRA operator [13].

Clojush also implements a method of automatic simplification, which takes a program and converts it into a smaller, semantically equivalent program. This process uses hill-climbing to remove instructions and code blocks from the program, checking at each step that the resulting program produces the same error vector as the original program [14]. This can dramatically simplify programs, reducing, for example, one program from 194 instructions down to nine instructions.

Our goal here is to give a deep analysis of a single run of PushGP, exploring and analyzing many of the programs, selections, and variations that make up this run. We chose to analyze a run on the Replace Space With Newline (RSWN) problem, taken from a recent general program synthesis benchmark suite [6]. In this problem, a program is given a string as input and should perform two tasks: first, it must print the result of replacing each space in the input with a newline character, and second, it must functionally return the number of non-whitespace characters in the input by leaving that value on top of the *integer* stack.

To store and process our ancestry data we used the Titan graph database along with the Gremlin shell and the Apache Tinkerpop query language.[3] This allowed us to store information about nodes (individuals), such as genomes and error vectors, and edges (relations) such as parent–child relationships. The graph database tools then make it easy to trace lineages and extract the subgraphs visualized in the next section. For these visualizations we used the Graphviz `dot` graph layout tool.[4]

5.3 Ancestry Graphs

The run we analyze here used a population size of 1000. This particular run found a solution after 20 generations, so we stored a total of 21,000 individuals in the graph database for this run. There were thirteen different "winning" individuals in that final generation, each of which had zero error on all of the 200 training cases.

In this section we describe two techniques for extracting and visualizing aspects of the run. The first is the ancestry tree, which contains of every ancestor (e.g., parents, grandparents, etc.) of any individual who found a solution. The second is the genetic ancestry tree, which is the subset of the ancestry tree limited to just those individuals that contributed at least one instruction to a particular successful individual.

5.3.1 Full Ancestry Graph

Figure 5.1 shows the full ancestry tree of the 13 successful individuals in this run. Each individual is represented with a rectangle containing an identifier of the form

[3]http://thinkaurelius.github.io/titan/ and https://tinkerpop.apache.org/.

[4]http://www.graphviz.org/.

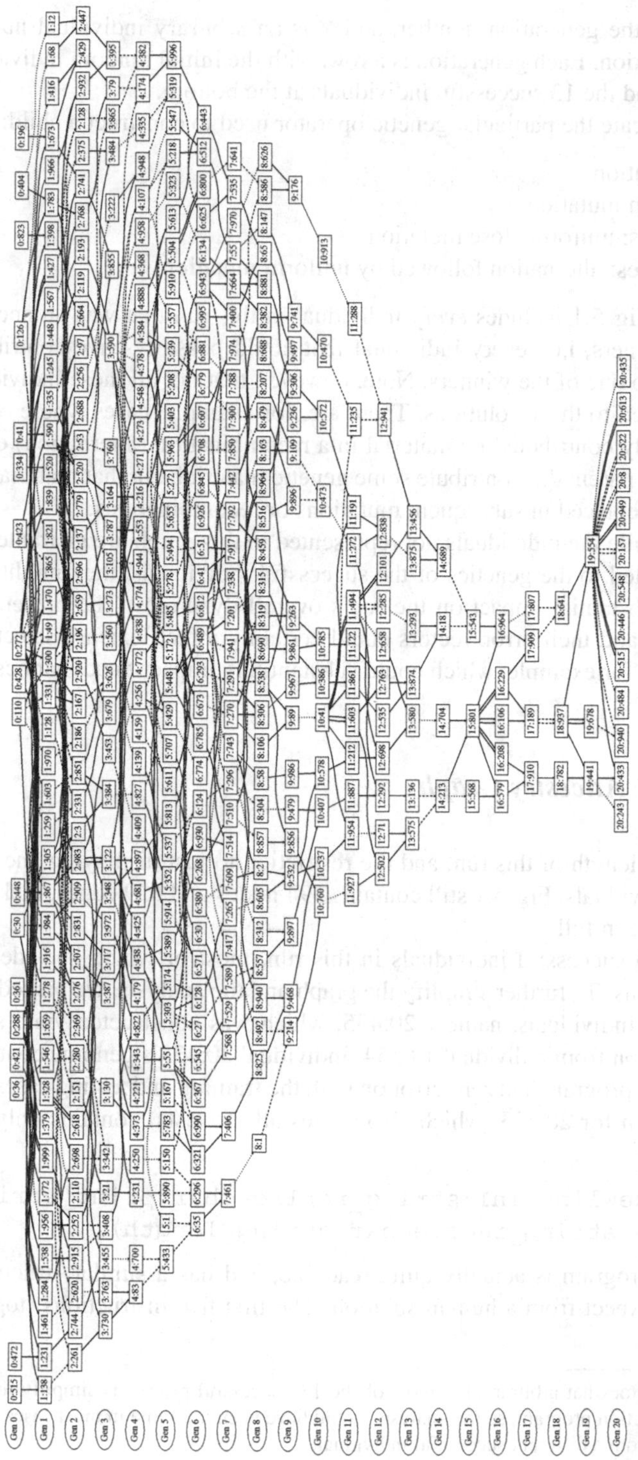

Fig. 5.1 The full ancestry graph containing all the ancestors of the 13 successful individuals. Individuals are represened with boxes, with the each generation being a row; the top row in the initial random population, and the bottom contains the 13 successful individuals. Edges represent parent–child relationships; see the text for descriptions of the meaning of the particular edge decorations

`X:Y`, where `X` is the generation number, and `Y` is an arbitrary individual number within that generation. Each generation is a row, with the initial random individuals being at the top and the 13 successful individuals at the bottom.

The edges indicate the particular genetic operator used to construct a child:

- Dashed: alternation
- Dotted: uniform mutation
- Thin black lines: uniform close mutation
- Thick black lines: alternation followed by uniform mutation

The graph in Fig. 5.1 includes *every* individual in this run that was an ancestor of one of the winners, i.e., every individual that could possibly have contributed genetic material to one of the winners. Note, however, that not all these individuals actually contributed to those solutions. There are, for example, cases where one of the parents actually contributed no material in a recombination (alternation) event, and cases where a parent did contribute some genetic material, but that material was later removed or replaced in subsequent mutations or recombinations.

Conversely, while the individuals not represented in this graph are guaranteed to have not contributed to the genetics of the successful individuals, they might have still had some substantial impact on the run's overall dynamics. The presence of those individuals and their error vectors could certainly affect lexicase selection's choice of parents, for example, which could substantially impact the dynamics.

5.3.2 Genetic Ancestry Graph

Despite the short length of this run, and the restriction to just displaying ancestors of successful individuals, Fig. 5.1 still contains 394 nodes and 629 edges, making it difficult to analyze in full.

There were 13 successful individuals in this run, most of which had identical simplified programs. To further simplify the graph and the analysis, we picked[5] one of the successful individuals, namely 20:435, which was constructed via a single instruction mutation from individual 19:554. Individual 20:435's genome contained 194 genes, and its program had zero error on both the training and testing cases. The simplified program for 20:435 (which also passes all the tests) contains only nine instructions:

```
(\space \newline in1 string_replacechar print_string
in1 \space string_removechar string_length)
```

This simplified program is actually quite readable, and has a similar structure to what me might expect from a human solution. The first five instructions (together

[5]This choice was somewhat arbitrary, but most of the 13 successful programs simplify down to the same nine instruction program, so the analysis would have been the same in most cases even if we'd worked back from a different successful individual.

on the first line) replace all the spaces in the input string with newlines (using the `string_replacechar` instruction) and print the resulting string, thereby solving half the problem. The next four instructions (on the second line) remove all the spaces from a fresh copy of the input string, compute the length and leave that on the `:integer` stack as the "returned" result.

To simplify the graph in Fig. 5.1, we extracted the subgraph containing only those individuals that contributed at least one of these nine *key instructions* to individual 20:435; see Fig. 5.2. Starting from 20:435 we traced backwards through it's ancestors, tracking where the nine key instructions came from. In doing so we found all of the members in the full ancestry graph that contributed these important instructions. and then extracted the genetic ancestry subgraph containing only these individuals. By cutting down on the number of individuals displayed we have a much more readable and focused visualization of this important ancestry information.

The genetic ancestry graph in Fig. 5.2 uses the same basic display of node and edge information as the full ancestry graph. There are, however, several additions that indicate how the nine key instructions flowed through the ancestry. First, we decorated nodes with boxes showing which of the nine key instructions were present in that individual. In most cases undecorated nodes contain the same instructions as the most recent labeled ancestor; the exceptions to this are individuals 16:964, 17:909, and 18:641, each of which contribute just the `string_length` instruction inherited from 15:543. Next we added a thicker border to certain nodes, to indicate the introduction or combination of key instructions via either mutation or crossover. Individual 1:590 is highlighted, for example, because the key `print_string` instruction was introduced their via uniform mutation, and individual 10:473 is highlighted because the alternation of 9:109 and 9:896 brought together an `in1` instruction from 9:109 with the four printing instructions from 9:896. Lastly we used a grayscale color gradient to indicate which and how many of the instructions were present in an individual. The earlier of the nine key instructions are assigned lighter colors in the gradient, and the later instructions are assigned darker gradient colors. So individuals like 7:338 have a fairly "flat" gray because they contributed just a single instruction from near the middle of the program, where 19:554 has a strong gradient because it contributed all nine of the instructions.

The other important extension to the graph in Fig. 5.2 is that we labeled each edge with the Damerau–Levenshtein distance (DL-distance) between the genome vectors for each parent–child pair. The genome vectors were generated by concatenating the `:instruction`[6] and `:close` fields from each gene into a single sequence. As an example, the genome of successful individual 20:435 starts

```
{:instruction boolean_and, :close 0}
{:instruction boolean_shove, :close 0}
{:instruction exec_do*count, :close 0}
```

[6]Instructions were treated as atomic symbols when computing the Damerau–Levenshtein distances; swapping a `exec_if` with a `print_string` would only add a distance of 1.

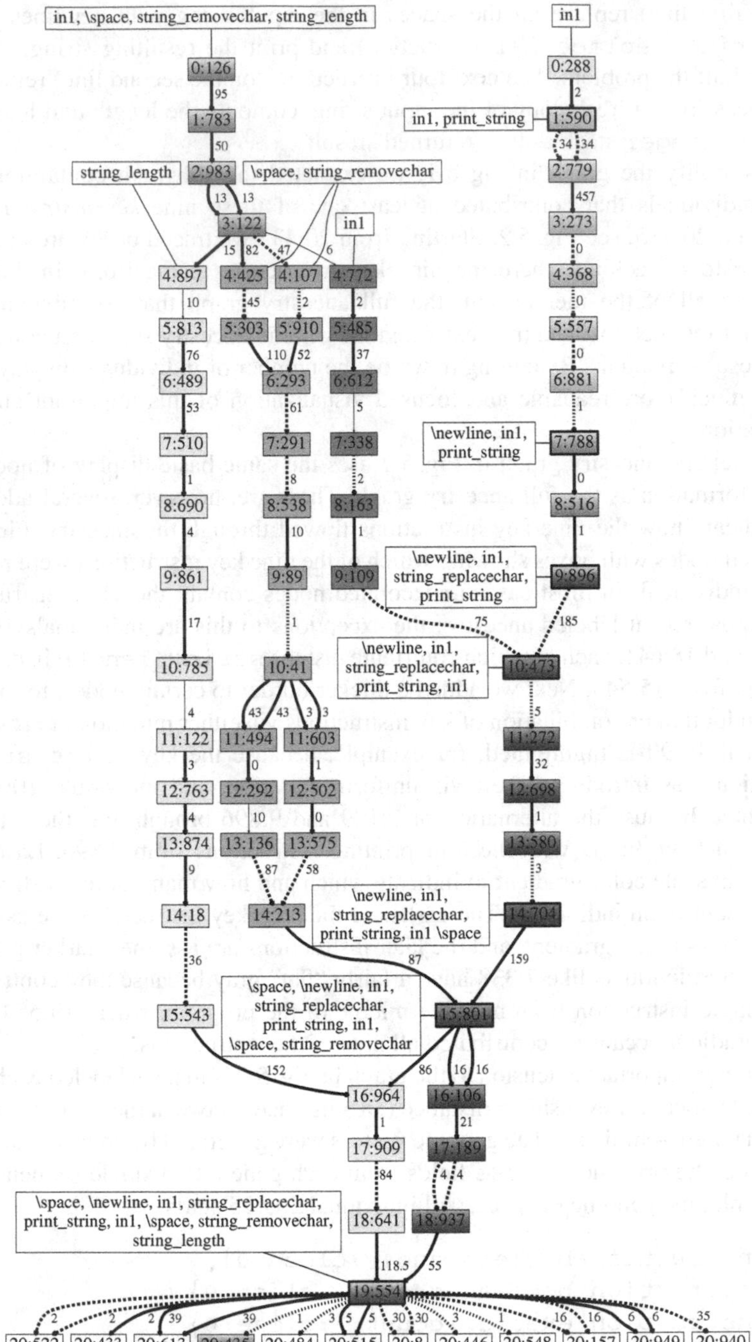

Fig. 5.2 The genetic ancestry version of the run's full ancestry graph

```
{:instruction exec_swap, :close 0}
{:instruction integer_empty, :close 0}
...
```

making the associated genome vector

```
boolean_and 0 boolean_shove 0 exec_do*count 0
exec_swap 0 integer_empty 0 ...
```

The Damerau–Levenshtein distance provides a succinct way to see when an individual has received a large amount of genetic material from its parents. It also allows us to easily identify alternation events that have mutation-like behavior, where there is only a small difference between to genome of one of the parents and the child.

5.4 The (Successful) End and How We Got There

As discussed earlier, individual 20:435's program simplifies down to just nine instructions:

```
(\space \newline in1 string_replacechar print_string
in1 \space string_removechar string_length)
```

where the first five instructions (the first line) handle the printing part of the Replace Space With Newline problems, and the next four instructions (the second line) handle the requirement that the program returns the number of non-space characters.

In this section we trace the origin of each of these nine instructions, going back to their introduction either via a mutation or as an element of one of the initial, random programs in the first generation. It's clear that each of these was "necessary" for the construction of this particular solution, so knowing where they all came from and how they came together should give us a valuable sense of the dynamics of this run. It's important to realize, however, that this will never be the whole story. Push instructions and values can play an important role in subtle ways, e.g., as spacers on stacks that when "counting" is implemented with a stack depth command. Removal of instructions can also be important. One key step in this run, for example, is the removal in the construction of 15:801 of an extraneous `print_newline` present in 14:704; the presence of this instruction caused the printed output to always have an error of one, and its removal changed all of the 100 "printing" errors from 1 to 0. All that said, however, we need some way to limit the number of individuals and events to analyze, so here will focus on the how those nine instructions trace through the ancestry.

It's also important to note that we didn't actually collect enough information to always say for *certain* where an instruction came from in a recombination event. There are numerous copies of instructions like `in1` in most of the genomes, for example, and in principle any of them in a parent could be the source of an `in1` in a child. In practice, however, there are constrains of location and order that typically allowed us to identify a single, unique source. There were a few places, however, where judgement calls were made. In future work we're going to explore attaching unique IDs to each gene and track not just parent–child relationships, but also source-destination relationships among genes, as this will give us certainty about the sources of genes, and allow us to automate more of the analysis, all at the expense of larger databases.

Returning to the specific program, it turns out that the evolution of the first five instructions, those handling the printing part of Replace Space With Newline, is largely independent of the evolution of the last four instructions, which handle the return part of the problem. The first five instructions, for example, all appear early in the genome for 20:435, between gene 9 and gene 24, while the last four all appear much later in the genome, between gene 107 and 175. As a consequence we'll trace these two groups one at a time, then discuss the "end game" after those two groups of instructions are brought together in individual 19:554.

5.4.1 *Printing: The First Five Instructions*

The ancestry of the five instructions that solve the printing test cases is fairly straightforward, and involves far more mutation than we expected. Unlike the return case instructions described below (Sect. 5.4.2), here there is a clear linear path for these instructions. They are introduced over time, and they are never split apart into branches to be recombined later.

Starting at the top-right of Fig. 5.2 with individual 0:288, only one of the key instructions, `in1`, was present in that individual from the initial randomly generated population. The other four of the five key printing instructions were introduced over time through a series of uniform mutations. The second of these five key instructions was introduced in individual 1:590 via a point mutation converting a piece of 0:288's genome into a `print_string` instruction. These two instructions are passed along this branch until they are joined by the next important instruction, `\newline`, introduced in 7:788 again via uniform mutation. After descending another two generations these three instructions were joined by `string_replacechar` in 9:896 via yet another uniform mutation. The final of these five key instructions, `\space` was added in generation 14 via a uniform mutation of 13:580 into 14:704, thus completing the five key printing instructions. These five instructions were then passed down as a group through 15:801 to the winners.

The impact of these instruction additions can be seen in the individuals' error vectors, as each addition was accompanied by a shift in the printing test cases. Sometimes the effect was minimal, with both small positive and small negative changes in the errors for different test cases, while other times the change led to dramatic improvements. As an example of a dramatic improvement, when `print_string` was introduced into 1:590, for example, the error on nearly all of the 100 printing test improved, with only a few showing an increased error; the total error for 1:590 across all 200 test cases was 492, where it's parents (0:41[7] and 0:288) has total errors of 1594 and 1154 respectively. Later, in the creation of 14:704 via mutation from 13:580 all of the printing test scores became 1. This was in general an *vital* step, but did lead to an increased error on a few tests that had passed with no error in 13:580; the total error of 14:704 was 922, where the total error of 13:580 was a slightly worse 1125.

5.4.2 *Returning: The Last Four Instructions*

The last four "returning" instructions were present in the right order in the very first generation, in individual 0:126. The first of these instructions (`in1`) was on gene 75 of the genome, the next two (`\space` and `string_removechar`) were on genes 89 and 94, and then the final instruction (`string_length`) was on line 141 (out of a total of 161 genes in that initial genome). Despite the fact that this individual had "all the right stuff", it's error vector had very few zeros, i.e., it was rarely correct, highlighting the fact that the presence or absence of other instructions can profoundly impact a program's behavior. 0:126 was, however, quite good for a randomly generated program, with all it's errors being under 20, and most being in the single digits. It was selected 45 times to be a parent, making it the seventh most selected parent in the initial generation, and one of only 48 individuals in the initial generation that received any selections. (The most selected parent in that generation, 0:272, was selected 762 times, but ultimately contributed no genes to the winning individual and therefore is not shown in the graph.)

Those four instructions were passed on as a group, with nearly the same relative positions in the genomes, from 0:126 through 1:783 and 2:983 to 3:122 (see Fig. 5.2). 3:122 was the third most selected individual in its generation and had 100 children, several of which went on to carry one or more of these four instructions forward to individual 19:554 when they were finally reunited in the positions that would ultimately lead to success. In particular there were three distinct branches coming from 3:122, each of which will be discussed below.

5.4.2.1 Branch 4:772 and the Carriers of `in1`

Individual 4:772 inherited the copy of the first instruction, `in1`, that would ultimately form part of the solution. This was transmitted down to 9:109 where it was recombined with 9:896 which, as mentioned above in Sect. 5.4.1, carried all but one of the first five "printing" instructions.

[7] Individual 0:41 isn't shown in Fig. 5.2 since it didn't contribute any of the nine key instructions to 1:590 or, ultimately, 20:435.

This recombination led to individual 10:473, which then had four of the five "printing" instructions, as well as the `in1` that would be the first of the four "returning" instructions. These five instructions were then passed down to 14:704, along with the `\space` introduced by mutation in 14:704. 14:704 was one of the parents of 15:801, a recombination which will be described in the discussion of the next branch.

5.4.2.2 Branches 4:425, 4:107, and Multiple Blocks

3:122 contained a block of 25 genes that contain the two middle instructions in the "returning" code, `\space` and `string_removechar`. This block was replicated in both 4:425 and 4:107, and then passed, respectively, to 5:303 and 5:910. 5:303 and 5:910 then recombined to create 6:293, which ended up having two complete copies of this block of genes.

These two copies of this block were then copied from 7:291 down through 10:41, to both 13:136 and 13:575. When these recombined to form 14:213 we ended up with *three* near copies of the block. These blocks were no longer identical due to small changes caused by earlier genetic operations, but each block still contained over 20 genes shared, including the two key instructions, `\space` and `string_removechar`, still four instructions apart.

All three of these blocks (and their three copies of these two "final" instructions) were passed on to 15:801 in the recombination of 14:213 and 14:704. 14:704 also bequeathed to 15:801 all of its six "final" instructions, meaning that 15:801 had all but one of those nine instructions, missing only the final `string_length`.

5.4.2.3 Branch 4:897 and the Carriers of `string_length`

4:897 and its descendants carried the copy of the last instruction, `string_length`, that would ultimately form part of the solution. This was transmitted all the way down to 18:641 without any significant interactions with other "final instructions", as is reflected in the almost entirely linear ancestry from 4:897 to 18:641 in Fig. 5.2.

There was one potentially interesting interaction with the other branches, when 15:543 combined with 15:801 to create 16:964. In this recombination event, however, 15:801 did not contribute any of the nine key instructions to 16:964. 15:543, on the other hand, transmitted the crucial missing `string_length` gene that had been passed down since our starting random generation, and which went on to be part of the solution in 20:435.

5.4.3 *From 19:554 to the End, and the Final Adjustments*

19:554 was the result of a recombination of 18:641 and 18:937, which finally brought together all nine of the "final" instructions. 18:937 contributed the first eight instructions, and 18:641 contributed the final `string_length` instruction. Individual 19:554 didn't *quite* solve the problem, however, it did have three "return" test cases with error 1. These three test cases turned out to be the only cases with an input of a single character.

These errors were fairly easy to rectify, however, as evidenced by the fact that 12 of 19:554's 747 offspring (or 1.6%) were indeed successful. Two of these successful children (20:435 and 20:548) were the result of mutating a *single* instruction. The key change brought about by the mutation that led to 20:435 caused an instance of the instruction `string_butlast` to not operate. In 19:554 this `string_butlast` was incorrectly removing the one and only character from the input string when the input consisted of a single character string, so the suppression of that instruction led to a perfect solution.

5.5 Discussion

The trace in Sect. 5.4 provides a sense of where all the key instructions came from, and indicates several of the key moments in the evolutionary process. In this section we'll provide some summary information as well as highlighting both some general patterns and a few important events.

Table 5.1 enumerates the number and proportions of individuals constructed via the four genetic operators, first across the entire run (so all of the 20,000 individuals generated after the initial random population), then for the ancestry graph in Fig. 5.1 (394 total nodes, 376 constructed after the initial generation), and finally for the genetic ancestry graph in Fig. 5.2 (62 total nodes, 60 constructed after the initial generation). The percentages in the "Entire run" column match the settings in the run configuration, which specified using alternation followed by uniform mutation 50% of the time, alternation alone 20% of the time, uniform mutation 20% of the time, and uniform close mutation the remaining 10% of the time. The other two columns have similar percentages, suggesting that there wasn't a large skew away from those parameter values, and that none of the genetic operators were particularly over- or under-represented in the ancestry graphs.

While there are numerous alternations in the genetic ancestry graph, it's worth noting that many of the DL-distances (the edge labels in Fig. 5.2) are fairly small, even when alternation was involved, as can be seen in Fig. 5.3. Of the 53 alternations in the genetic ancestry graph (ignoring those leading to a successful individual in generation 20), 21 had DL-distances of 10 or less, six had DL-distances of just 1, and five had DL-distances of 0 (the child was an exact copy of a parent). One might assume that this is partly due to the six self-cross alternations, where the

Table 5.1 The numbers and proportions of individuals constructed using the different genetic operators

Genetic operator	Entire run	Full ancestry	Genetic ancestry
Alternation + uniform mutation	9985 (50%)	186 (49%)	39 (54%)
Alternation	4001 (20%)	67 (18%)	17 (24%)
Uniform mutation	4026 (20%)	83 (22%)	11 (15%)
Uniform close mutation	1988 (10%)	40 (11%)	5 (7%)

Total percentages might not equal 100% due to rounding

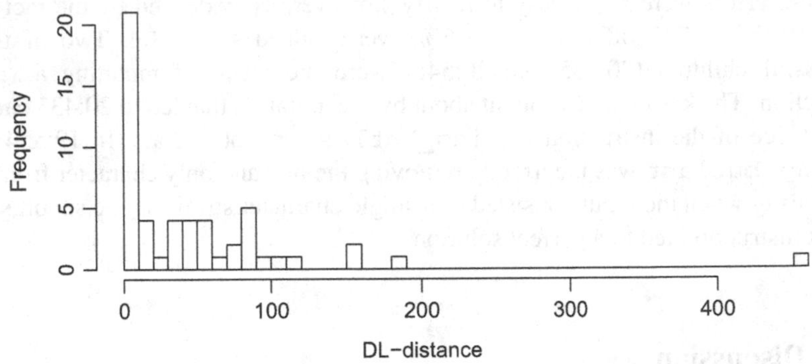

Fig. 5.3 The distribution of DL-distances for all the alternation events in the genetic ancestry graph (Fig. 5.2) whether or not they were followed by uniform mutation. This does not include the alternations leading to successful individuals in the final generation since those were almost all self-crosses, which skew towards smaller DL-distances

same individual served as both parents, such as individual 16:106 having 15:801 as both of its parents.[8] In fact, however, most of the self-crosses in the genetic ancestry graph had higher than median DL-distances.

These very small DL-distances mean that many of the alternations were effectively acting as mutation-like events. The steps from individual 15:801 to 18:937, for example, are all alternations (possibly followed by mutations), but in fact almost every change in that sequence was due to gene deletions or duplications in those alternation events. There were three mutated genes in that sequence of steps, along with 12 deleted genes and the duplication of a block of seven genes.

Since the genetic ancestry graph (and thus the data in Fig. 5.3) only includes individuals that actually contributed one of the nine key instructions, in many cases the second parent in alternation events isn't included; these DL-distances are in general higher than those listed. This isn't surprising, as a parent with a small DL-distance is very similar to the child, and thus likely to have contributed most of the important genetic material. There are, however, a few exceptions to this pattern. Perhaps the most extreme is in the creation of individual 3:273 via alternation

[8]These self-crosses are likely a result of hyperselection events due to lexicase selection [8].

between 2:659 and 2:779. Individual 2:659 is not included in the genetic ancestry graph in Fig. 5.2 because it didn't contribute any of the nine key instructions to 3:273, whereas 2:779 contributed two such instructions. However the DL-distance between 3:273 and 2:779 was 457, which was much greater than the distance to 2:659, which was only 50. So despite being much more similar to 2:659 and getting most of its genetic material from that parent, the material that ultimately contributed to the solution all came from the other parent (2:779).

Not all alternation events in Fig. 5.2 could effectively be seen as mutation events, however. The construction of 15:801, for example, was in many ways what we imagine when we think about crossover events, combining significant genetic material and significant functionality from two different parents. It was also a key point in the run, as 15:801 was the first individual to be correct on all of the "printing" test cases, and it was also correct on 26 of the 100 "returning" test cases.

Individual 15:801 was created through the recombination of 14:704 and 14:213, via alternation followed by uniform mutation. Table 5.2 shows the simplified programs of both parents and the child, aligned to indicate where the various instructions likely came from. The key observation is that 15:801 received most of its initial genetic material from 14:704 (most of genes 1–6), followed by a large section (genes 7–35) taken almost entirely from 14:213's genome. Interestingly, the transition between 14:704 and 15:801 involved a simple but crucial change that fixed all the printing cases. 14:704 had an error of exactly 1 on all the printing cases due to an extra `print_newline` (line 37 in Table 5.2). In the recombination this gene wasn't passed on to 15:801, which led to a perfect score of 0 on all those test cases. The performance of 15:801 on the return test cases wasn't quite as strong as that of its other parent, 14:213, but was generally better than 14:704's performance on those test cases. 15:801 went on to receive a large number of selections (595) and being a parent of just over half the next generation (501 individuals).

5.6 Conclusions and Future Work

Here we traced through the genetic ancestry of a short, successful genetic programming run. While the run was short, it used an "industrial strength" PushGP system on a non-trivial problem that required the manipulation of strings and integers in multiple ways, and a combination of both printing and returning results. We used graph database tools to create ancestry and genetic ancestry graphs, which we were then able to use to visualize and analyze this run. The resulting graphs show the progression of the run and highlight important moments such as key recombination events, gene deletions and duplications, and the introduction of key instructions via mutation. By tracing through the genetic ancestry tree we were able to learn more about how both alternation and mutation played a role in finding a solution.

While we were able to do this for a small run, currently too much of the process is manual for this to scale to larger runs or multiple sets of runs. A key next step in further automating this kind of analysis is automating the process of comparing

Table 5.2 The details of the recombination event (alternation followed by uniform mutation) that created individual 15:801 (center) from parents 14:704 (left) and 14:213 (right) showing the *simplified* programs for those individuals (see Sect. 5.2)

14:704	15:801		14:213
	0		(in1
(\space	1	(\space	
\newline	2	\newline	
	3	exec_dup	
in1	4	in1	
string_replacechar	5	string_replacechar	
print_string	6	print_string	print_string
	7	exec_dup	exec_dup
	8		exec_s
	9		(exec_dup
	10		(exec_rot
	11	(string_eq	(string_eq
	12		string_fromboolean)
	13		char_eq
	14		(string_emptystring
	15		boolean_stackdepth
	16		in1
	17		integer_gt)
	18		string_emptystring
	19	\space	\space
	20	string_dup	string_dup
	21	string_removechar	string_removechar
	22	string_rot	
	23		boolean_pop
	24		in1
	25		string_butlast
	26		string_last
	27		string_parse_to_chars
	28		exec_when
	29		string_dup
	30		string_removechar
	31	string_last	string_last
	32	string_parse_to_chars	string_parse_to_chars
	33	string_rot)	string_rot)
	34	in1	in1)
	35	string_stackdepth)	string_stackdepth)
boolean_stackdepth	36		
print_newline)	37		

This shows that individual 15:801 was essentially constructed from a short prefix of 14:704 and a longer suffix of 14:213

individuals, especially at the genome level. Tracing each key instruction back through the ancestry graph can be complicated, in part because there are often many different instances of the instruction being traced; individual 19:554, for example, had four instances of `\space`, but only three of those were present in its simplified program, and only two went on to be part of the simplified successful program in 20:435. In this case we were able to deal with these problems by using contextual clues such as order in the genome and surrounding instructions, not unlike how biologists track gene sequences in organisms. To make this process more automatic and exact, however, we'll need to save additional information with the individual genes that allows us to know exactly where they came from.

It would also be valuable to improve our ability to understand and compare program behaviors. We can easily compare genomes and error vectors, and reasonably compare program *texts*, comparing program *behaviors* is much less straightforward. While the simplified program for individual 20:435 is quite short and understandable, the unsimplified program contains 195 instructions, which include a number of complex looping constructs. These are obviously not necessary for the semantics of the program, but they are present in the code that is being tested, and the genes that create those instructions are part of the genome that is being manipulated and inherited. And while those instructions might be removable from 20:435 at the end of the run, it's likely that many of those instructions played some meaningful role in an ancestor that contributed to that ancestor's selection.

Lastly the prevalence of numerous alternation events in the gene ancestry graph that turned out to be just gene deletions or duplications suggests that it might be valuable to include deletion and replication mutations as stand-alone operators, instead of requiring that such events occur via lucky alternations.

Acknowledgements Emma Sax, Laverne Schrock, and Leonid Scott helped with the initial computation and analyses of the differences between the parents and children discussed here. William Tozier provided a host of ideas and feedback all through the process, as did numerous members of the Hampshire College Computational Intelligence lab.

We are very grateful to all the participants in the 2016 Genetic Programming Theory and Practice (GPTP) workshop for their enthusiasm, ideas, and support. In particular we'd like to thank William Tozier, Stephan Winkler, and Wolfgang Banzhaf for their feedback on an earlier draft. Finally, thanks to the GPTP organizers; without their hard work none of those other valuable conversations would have occurred.

This material is based upon work supported by the National Science Foundation under Grants No. 1129139 and 1331283. Any opinions, findings, and conclusions or recommendations expressed in this publication are those of the authors and do not necessarily reflect the views of the National Science Foundation.

References

1. Burlacu, B., Affenzeller, M., Kommenda, M., Winkler, S., Kronberger, G.: Visualization of genetic lineages and inheritance information in genetic programming. In: GECCO '13 Companion: Proceeding of the Fifteenth Annual Conference Companion on Genetic and Evolutionary Computation Conference Companion, pp. 1351–1358. ACM, Amsterdam (2013)

2. Burlacu, B., Kommenda, M., Affenzeller, M.: Building blocks identification based on subtree sample counts for genetic programming. In: Asia-Pacific Conference on Computer Aided System Engineering (APCASE), 2015, pp. 152–157. IEEE, Piscataway (2015)
3. Burlacu, B., Affenzeller, M., Winkler, S., Kommenda, M., Kronberger, G.: Methods for genealogy and building block analysis in genetic programming. In: Computational Intelligence and Efficiency in Engineering Systems. Studies in Computational Intelligence, vol. 595, pp. 61–74. Springer, Cham (2015)
4. Burlacu, B., Affenzeller, M., Kommenda, M.: On the effectiveness of genetic operations in symbolic regression. In: Computer Aided Systems Theory–EUROCAST 2015, pp. 367–374. Springer, Cham (2015)
5. Donatucci, D., Dramdahl, M.K., McPhee, N.F.: Analysis of genetic programming ancestry using a graph database. In: Proceedings of the Midwest Instruction and Computing Symposium (2014). http://goo.gl/RZXY2U
6. Helmuth, T., Spector, L.: General program synthesis benchmark suite. In: Silva, S., Esparcia-Alcazar, A.I., Lopez-Ibanez, M., Mostaghim, S., Timmis, J., Zarges, C., Correia, L., Soule, T., Giacobini, M., Urbanowicz, R., Akimoto, Y., Glasmachers, T., Fernandez de Vega, F., Hoover, A., Larranaga, P., Soto, M., Cotta, C., Pereira, F.B., Handl, J., Koutnik, J., Gaspar-Cunha, A., Trautmann, H., Mouret, J.B., Risi, S., Costa, E., Schuetze, O., Krawiec, K., Moraglio, A., Miller, J.F., Widera, P., Cagnoni, S., Merelo, J., Hart, E., Trujillo, L., Kessentini, M., Ochoa, G., Chicano, F., Doerr, C. (eds.) GECCO '15: Proceedings of the 2015 Annual Conference on Genetic and Evolutionary Computation, pp. 1039–1046. ACM, Madrid (2015). https://doi.org/10.1145/2739480.2754769. http://doi.acm.org/10.1145/2739480.2754769
7. Helmuth, T., Spector, L., McPhee, N.F., Shanabrook, S.: Plush: linear genomes for pushgp. In: Genetic Programming Theory and Practice XIV. Genetic and Evolutionary Computation. Springer, Ann Arbor (2016)
8. Helmuth, T., McPhee, N.F., Spector, L.: The impact of hyperselection on lexicase selection. In: GECCO '16: Proceedings of the 2016 Genetic and Evolutionary Computation Conference (2016)
9. Kuber, K., Card, S.W., Mehrotra, K.G., Mohan, C.K.: Ancestral networks in evolutionary algorithms. In: Proceedings of the 2014 Conference Companion on Genetic and Evolutionary Computation Companion, pp. 115–116. ACM, New York (2014)
10. McPhee, N.F., Donatucci, D., Helmuth, T.: Using graph databases to explore genetic programming run dynamics. In: Riolo, R., Worzel, W.P., Kotanchek, M., Kordon, A. (eds.) Genetic Programming Theory and Practice XIII, Genetic and Evolutionary Computation. Springer, Ann Arbor (2015). https://doi.org/10.1007/978-3-319-34223-8. http://www.springer.com/us/book/9783319342214
11. Spector, L.: Assessment of problem modality by differential performance of lexicase selection in genetic programming: a preliminary report. In: McClymont, K., Keedwell, E. (eds.) 1st Workshop on Understanding Problems (GECCO-UP), pp. 401–408. ACM, Philadelphia (2012). https://doi.org/10.1145/2330784.2330846. http://hampshire.edu/lspector/pubs/wk09p4-spector.pdf
12. Spector, L., Robinson, A.: Genetic programming and autoconstructive evolution with the push programming language. Genet. Program Evolvable Mach. **3**(1), 7–40 (2002). https://doi.org/10.1023/A:1014538503543. http://hampshire.edu/lspector/pubs/push-gpem-final.pdf
13. Spector, L., Helmuth, T.: Uniform linear transformation with repair and alternation in genetic programming. In: Riolo, R., Moore, J.H., Kotanchek, M. (eds.) Genetic Programming Theory and Practice XI. Genetic and Evolutionary Computation, chap. 8, pp. 137–153. Springer, Ann Arbor (2013). https://doi.org/10.1007/978-1-4939-0375-7_8
14. Spector, L., Helmuth, T.: Effective simplification of evolved push programs using a simple, stochastic hill-climber. In: Igel, C., Arnold, D.V., Gagne, C., Popovici, E., Auger, A., Bacardit, J., Brockhoff, D., Cagnoni, S., Deb, K., Doerr, B., Foster, J., Glasmachers, T., Hart, E., Heywood, M.I., Iba, H., Jacob, C., Jansen, T., Jin, Y., Kessentini, M., Knowles, J.D., Langdon, W.B., Larranaga, P., Luke, S., Luque, G., McCall, J.A.W., Montes de Oca, M.A., Motsinger-Reif, A., Ong, Y.S., Palmer, M., Parsopoulos, K.E., Raidl, G., Risi, S., Ruhe, G., Schaul, T.,

Schmickl, T., Sendhoff, B., Stanley, K.O., Stuetzle, T., Thierens, D., Togelius, J., Witt, C., Zarges, C. (eds.) GECCO Comp '14: Proceedings of the 2014 Conference Companion on Genetic and Evolutionary Computation Companion, pp. 147–148. ACM, Vancouver (2014). https://doi.org/10.1145/2598394.2598414. http://doi.acm.org/10.1145/2598394.2598414

15. Spector, L., Klein, J., Keijzer, M.: The push3 execution stack and the evolution of control. In: Beyer, H.G., O'Reilly, U.M., Arnold, D.V., Banzhaf, W., Blum, C., Bonabeau, E.W., Cantu-Paz, E., Dasgupta, D., Deb, K., Foster, J.A., de Jong, E.D., Lipson, H., Llora, X., Mancoridis, S., Pelikan, M., Raidl, G.R., Soule, T., Tyrrell, A.M., Watson, J.P., Zitzler, E. (eds.) GECCO 2005: Proceedings of the 2005 Conference on Genetic and Evolutionary Computation, vol. 2, pp. 1689–1696. ACM Press, Washington DC (2005). https://doi.org/10.1145/1068009.1068292. http://www.cs.bham.ac.uk/~wbl/biblio/gecco2005/docs/p1689.pdf

Chapter 6
Linear Genomes for Structured Programs

Thomas Helmuth, Lee Spector, Nicholas Freitag McPhee, and Saul Shanabrook

Abstract In most genetic programming systems, candidate solution programs themselves serve as genome upon which variation operators act. However, because of the hierarchical structure of computer programs and the syntactic constraints that they must obey, it is difficult to implement variation operators that affect different parts of programs with uniform probability. This lack of uniformity can have detrimental effects on evolutionary search, such as increases in code bloat. In prior work, structured programs were linearized prior to variation in order to facilitate uniform variation. However, this necessitated syntactic repair after variation, which reintroduced non-uniformities. In this chapter we describe a new approach that uses linear genomes that are translated into hierarchical programs for execution. We present the new approach in detail and show how it facilitates both uniform variation and the evolution of programs with meaningful structure.

Keywords Genetic programming · Uniform variation · Linear genome · Push language · Plush genome · Representation scheme

---

T. Helmuth (✉)
Computer Science, Washington and Lee University, Lexington, VA, USA
e-mail: helmutht@wlu.edu

L. Spector
Cognitive Science, Hampshire College, Amherst, MA, USA
e-mail: lspector@hampshire.edu

N. F. McPhee
Division of Science and Mathematics, University of Minnesota, Morris, MN, USA
e-mail: mcphee@morris.umn.edu

S. Shanabrook
Computer Science, University of Massachusetts, Amherst, MA, USA

R. Riolo et al. (eds.), *Genetic Programming Theory and Practice XIV*, Genetic and Evolutionary Computation,
https://doi.org/10.1007/978-3-319-97088-2_6

6.1 Introduction

In traditional tree-based genetic programming, genetic operators such as subtree crossover and subtree mutation exhibit biases as to how likely it is for any given component of a parent program to be transferred to the resulting child. These biases make these genetic operators, and indeed any genetic operators defined over tree structures, decidedly nonuniform. What is meant by "uniform" in this context? In prior work [18] we defined uniformity in terms of two desiderata for variation operators:

- "...that the probability of an inherited program component being modified during inheritance is independent of the size and shape of the parent programs beyond the component in question...," and
- "...that pairs of parents are combined in ways that allow arbitrary combinations of components from each parent to appear in the child."

Genetic programming's most common program representation leverages the relative simplicity of Lisp symbolic expressions, which can express richly structured programs despite having few syntactic constraints in comparison to other common programming languages [4]. Hierarchical symbolic expressions, represented by tree structures, simplify the implementation of genetic operators that produce syntactically valid children from syntactically valid parents, using processes of subexpression replacement and exchange. However, the widely-used operators based on these processes do not meet our definitions for uniformity. For example, in standard subtree mutation a single subexpression is chosen and replaced, making the chance of replacing each subexpression inversely proportional to the size of the overall program. So standard mutation violates the first uniformity desideratum. In standard crossover a single subexpression is replaced by a subexpression from the other parent, restricting the ways in which the components of the parents can be combined in children. Thus standard crossover violates the second uniformity desideratum.

Why do these deviations from uniformity matter? One issue is that the any particular subexpression is more likely to survive without modification if it is embedded within a large program rather than a small program. This can bias survival of important subexpressions toward larger programs, irrespective of fitness, leading to "code bloat" [9]. Another issue, related to the second uniformity desideratum, is that standard crossover does not permit complementary parts of two parents to be combined in their child, unless all of the needed parts from one parent are segregated in a single subexpression.

Several researchers have previously noted these and related issues and have attempted to address them through modification of the standard genetic operators. For example by making the probabilities of chosing a subexpressions dependent on its size [2, 4], adjusting the number of replacements or exchanges [23], and by restricting exchanges to pairs of expressions with specified properties [12, 14, 15]. As detailed in [18], none of these methods meet our definition of uniformity, limited in principle by the nested structure of the programs that are being modified.

Genetic programming systems which use linear program representations are immune to most of the problems raised here, because uniform genetic operators can be straightforwardly applied on linear sequences [11]. Much of the prior work in linear genetic programming is focused on programs expressed in a low-level language with few control and data structures. Here we aim to provide uniform variation for programs in a language that can support arbitrary control and data structures and for which program structure is therefore likely to be more important. An existing framework in which linear genomes can indeed be used to evolve highly structured programs is "grammatical evolution," in which the genes on linear genomes are used as indices into grammars that can express arbitrary languages [16]. While uniform genetic operators can indeed be used on these genomes, the effects that small changes to genomes have on the expressed programs are often quite large, so that uniformity at the level of genomes is unlikely to translate into uniformity at the level of programs.

In earlier work, we sought to achieve greater uniformity by treating programs as linear sequences only during variation [18]. Our ULTRA (Uniform Linear Transformation with Repair and Alternation) operator first translates hierarchically structured programs into linear sequences, with parentheses replaced by independent tokens. It then applies uniform mutation and alternation (a form of multipoint crossover) to the linear sequences. Finally, it translates the resulting linear sequences back into hierarchical programs. Because the tokens for parentheses may have become imbalanced during uniform variation, a repair step is required to rebalance them. While the prior work demonstrated that ULTRA had several desirable properties, the artifacts produced by the repair step were themselves non-uniform and biased the shape of evolving programs in peculiar ways. This nonuniformity was one motivation for the work presented here.

Another motivation for the work described below was the fact that while ULTRA supports reasonably-uniform variation of structured programs, it does nothing to produce structure where it is most likely to be useful, in the context of instructions that make use of structure, such as conditional branches or loops.

The idea for the alternative approach that we present here arose when considering the problems raised above in the context of independent work that we were conducting in which "epigenetic" markers were added to instructions and literals in linear programs in order to turn those genes on or off [5–7]. We realized we could use similar epigenetic markers to specify the hierarchical structure for programs that are "expressed" from linear genomes. This allows us to perform uniform genetic operators on linear genomes and only express them as hierarchical programs for fitness testing. Furthermore, we specify that opening parentheses are automatically inserted following structure-dependent instructions during translation and use epigenetic markers to indicate where closing parentheses should occur.

Thus we make it more likely that parenthesized, hierarchical structures appear in programs next to instructions that can make use of them.

We note that the use of the term "epigenetic" for these markers is most appropriate when they can change not only during reproduction, but also in their problem environments. While we do not describe such processes here, we have used these markers in this way in the past [7]. In that work, we used hill climbing to modify the epigenetic markers of an individual if those modifications improve its fitness. While this effort did not produce significantly better results, using similar mutations to turn off or on genes in newly created children, and allowing selection pressure to sort out the changes, did produce impressively better results on 2 out of 5 problems. So, even though we do not explore changing epigenetic markers during the "lifetimes" of the programs in the present work, the markers that we use do enable such modifications; additionally, because of the ways that these markers are attached to instruction and literal "genes," we think that the use of the label "epigenetic" is reasonable in this context.

In the remainder of this chapter we first provide a brief description of Push the programming language, which is expressed from our linear genomes. We then describe our new linear genome representation, which we call "Plush" (where the "l" is for "linear"), in detail. This description is followed by experimental results that demonstrate the ways in which Plush facilitates program structure and the efficacy of various uniform genetic operators.

6.2 Push and PushGP

The Push programming language was developed specifically to serve as the target language for program evolution in genetic programming and related program synthesis methods [19–21]. Push is a postfix, stack-based language, which is similar in some respects to others that have been used for genetic programming [13]. When a Push program is executed, literals are pushed onto data stacks and instructions act on data that is on the stacks. Among the types of data stored on stacks and manipulated by instructions is `code`, which permits the expression of complex control structures via code manipulation. Program execution is implemented through the decomposition of programs and the processing of their instructions and literals on a special stack that contains code, the `exec` stack. Because all instructions take their arguments from appropriately typed stacks, and because of the Push convention that instructions finding insufficient data on the relevant stacks act as `no-ops`, instructions and literals of any types can be interleaved in arbitrary ways without risk of type errors.

Like Lisp symbolic expressions, Push programs may be hierarchically structured with parentheses, and this structure has consequences for program execution when code-manipulation instructions are used. Unlike symbolic expressions, parenthesized code blocks may appear anywhere in a Push program, and their presence or absence does not change the syntactic validity of the program. For example, the `exec_if` instruction will execute one of the top two items on the `exec` stack, depending on the value on top of the `boolean` stack, and discard the other. Those items serve as the conditional execution branches of the if statement, and either may be a single instruction or a code block containing any number of grouped instructions. For example, in the program:

```
(arg1 exec_if (4 5 integer_add) 7)
```

if `arg1` is *true*, `7` (as the "else" clause) will be removed, and the block of code `(4 5 integer_add)` will remain on the `exec` stack. If `arg1` is *false*, the "then" clause `(4 5 integer_add)` will be popped and `7` will remain on the `exec` stack.

In early versions of PushGP, the parenthetic structure of programs also affected the ways that genetic operators operated on programs. The genetic operators in these versions of PushGP were intentionally similar to those used in traditional, Lisp symbolic expression genetic programming systems; mutation involved replacing subexpressions with new subexpressions, while crossover involved the exchange of subexpressions across programs. This facilitated comparisons between the different program representations, and the translation of ideas from one project to another. But, these subtree-based operators lacked uniformity for the same reasons traditional tree-based operators do.

6.3 Plush

Plush[1] *genomes* provide an alternative representation for Push programs, storing programs in linear sequences that enable simple uniform genetic variations. There is a many to one mapping from Plush genomes to Push programs, as described by the translation process below.

One of the goals of introducing Plush genomes is to ensure that every argument taken from the `exec` stack for use by an instruction consists of a parenthesized block of code. In prior work, when evolving Push programs as genomes, we would often see `exec` stack manipulating instructions taking single instructions as arguments, instead of blocks of code. By ensuring that such instructions are followed by code blocks, we hope to encourage more modular programs that can make better use of `exec` stack arguments.

The many Push instructions that take arguments from the `exec` stack include instructions for looping, conditional execution, and other program manipulation. For example, the instruction `exec_if` requires two `exec` stack arguments, one to execute if the condition is true and the other to execute if the condition is false. The instruction `exec_do*times` needs one `exec` stack argument, which is executed repeatedly in a loop. In the program in Sect. 6.2, the `exec_if` instruction takes two

[1] Linear **Push**.

arguments from the exec stack: (4 5 integer_add) and 7. The first is a block of code, and the second is a single instruction (in this case, an integer literal). Note that blocks of code can contain zero or more instructions, so that the above program could be replaced with a functionally equivalent one where each of exec_if's arguments is a block of code:

```
(arg1 exec_if (4 5 integer_add) (7))
```

6.3.1 Structure

Plush genomes are linear sequences of *gene maps*, each of which contains, at minimum, an *instruction* and an epigenetic *close count* marker used to determine the placement of parentheses during translation. For example, below is a very simple Plush genome that encodes the Push program (1 2 integer_add):

```
[ {:instruction 1, :close 0}
  {:instruction 2, :close 0}
  {:instruction integer_add, :close 0} ]
```

Plush gene maps can optionally contain other epigenetic markers. For example, the :silent marker contains a boolean that indicates whether the instruction should be included in the translated program. We have not yet added other epigenetic markers to Plush, but could imagine others being useful.

6.3.2 Translation

The process of translating a linear Plush genome into a syntactically valid, hierarchical Push program (which is a tree) is for the most part a depth-first construction of that program tree. The Plush genome is traversed linearly, adding each gene map's instruction to the end of the translated program. The hierarchical structure of the resulting program depends entirely on which of its instructions take arguments from the exec stack. Each instruction that does not take any arguments from the exec stack is simply appended to the growing program, and is not followed by a code block. An instruction that takes *X* arguments from the exec stack will be followed by *X* code blocks. After such instructions, further instructions will be added inside the opened code block, and will open nested code blocks when appropriate. Instructions only indicate where the code blocks open (i.e. they insert open parentheses); they do not describe where they should close (i.e. the location of the matching close parentheses).

As genes are translated from Plush into Push code, the values of the `:close` epigenetic markers determine where code blocks are closed. For every gene that is translated from Plush to Push: *after* the `:instruction` token has been added to growing Push program, and *after* a new block has been opened (if the instruction requires one), the `:close` marker is applied to the growing Push program. In particular, this number indicates the number of opened code blocks to close with closing parentheses. If the number is greater than the number of currently opened code blocks, all opened code blocks are closed. Note that if a code block is closed and the preceding instruction requires another code block (such as with `exec_if`, which requires two code blocks), one is immediately opened, which may be immediately closed if the `:close` marker is large enough. Finally, if the end of the genome is reached without closing all opened code blocks, the remaining blocks are automatically closed, including any blocks that still needed to be opened for instructions that take multiple `exec` stack arguments.

These *automatic code blocks* ensure that the hierarchical structure of a program has semantic meaning according to its instructions. It may help to think about a language such as Python or Java, in which it makes sense to block off a chunk of code following the start of a loop, but does not make semantic (or syntactic) sense to have a block of code follow a variable assignment. While such semantically-irrelevant code blocks are syntactically valid in Push, they have no affect on the semantics of the program.

6.3.3 Special Genes

The `:silent` epigenetic marker and two special instructions also affect the translation process:

- A Plush gene map with a `:silent` marker set to `true` is completely ignored and does not affect the growing Push program. Such genes have been "silenced."
- The `noop_open_paren` instruction immediately opens a new code block but adds no instruction to the Push program. No parenthesized branch is *ever* opened in the Push program unless the instruction takes one or more arguments from Push's `exec` stack or the instruction is `noop_open_paren`.
- The `noop_delete_prev_paren_pair` instruction restructures the Push program *without affecting the translation state in any other way*: it searches through the Push program until it finds the last block closed in translation, and "lifts" the contents of that block to the level of its parent in the program. For example, if the Push program is `[1 2 (3 4) 5 (6 *)]`, with the asterisk indicating where the next item would be added, inside a currently unclosed block, the result of applying this transformation would be `[1 2 3 4 5 (6 *)]`.

6.3.4 Example Translation

Here we give a brief example of a Plush genome and its corresponding Push program to illustrate the translation process. The genome:

```
[{:instruction exec_do*times :close 0}
 {:instruction 8 :close 0}
 {:instruction 11 :close 3}
 {:instruction integer_add :close 0 :silent true}
 {:instruction exec_if :close 1}
 {:instruction 17 :close 0}
 {:instruction noop_open_paren :close 0}
 {:instruction false :close 0}
 {:instruction code_quote :close 0}
 {:instruction float_mult :close 2}
 {:instruction exec_rot :close 0}
 {:instruction 34.44 :close 0}]
```

is translated into the Push program:

```
(exec_do*times (8 11) exec_if
  ()
  (17
   (false code_quote (float_mult))
   exec_rot (34.44) () ()))
```

The first instruction, `exec_do*times`, takes one argument from the `exec` stack, and therefore opens one code block. The next two instructions are added to this block, which is closed by the 3 `:close` marker. Note that while this marker says to close 3 code blocks, there is only one open block to close. This block is followed by a silenced gene containing the instruction `integer_add`, which is not added to the Push program.

Next, the `exec_if` instruction takes two arguments from the `exec` stack. Since the `:close` count of the `exec_if` instruction itself is 1, the first of those two blocks is immediately closed, and the second opened. Following 17 in the opened block, the `noop_open_paren` instruction opens a code block without adding an instruction to the Push program.

In the remainder of the Plush genome, the `code_quote` instruction takes one `exec` stack argument, which is closed along with another code block after the instruction `float_mult`. Finally, `exec_rot` opens three code blocks, none of which are closed by the end of the program. As such, these blocks are automatically closed at the end of the program.

6.4 Uniform Genetic Operators

One of the advantages of a linear genome representation is it allows us to use uniform genetic operators. This section describes in detail the genetic operators we use with Plush. For reference, Table 6.1 has the parameter settings related to genetic operators that we use in our experiments using the genetic operators described below, giving an idea of reasonable settings for these parameters; all indications thus far show these operators to be robust to changes in these parameter settings.

Uniform mutation modifies a single parent genome by changing each of its instructions with some probability, designated the *uniform mutation rate*. If an instruction in the genome is selected to be changed, we first check whether the instruction is a constant or a Push instruction. If it is an instruction, it is simply replaced by a random instruction from the instruction set. If it is a constant, there is a *constant tweak rate* probability of tweaking the constant; otherwise, it is replaced by a random instruction. The way in which a constant is tweaked depends on the type of the constant: integers and floats are perturbed by Gaussian noise with standard deviation of 1.0, strings have a 10% probability of replacing each character with a random character, and booleans are replaced by a random boolean.

While uniform mutation can change the instructions in a genome, it cannot affect the close epigenetic markers, and therefore cannot affect the structure of a program. We therefore created a *uniform close mutation* operator that takes a parent and alters its close markers. With *uniform close mutation rate* probability, it either increments or decrements the close marker associated with each instruction. The close marker cannot be decreased below 0, but has no upper bound. The probability of incrementing a close marker, as opposed to decrementing it, is given by the *close increment rate*; we typically keep this number less than 0.5, since otherwise we find that close markers tend to grow more than they shrink.

We use a crossover operator, *alternation*, heavily inspired by the ULTRA operator, which functioned on Push programs as genomes instead of linear genomes [18]. In alternation, both parents are traversed in parallel, copying instructions from one or the other into the child program. Before copying each gene, alternation has a small probability, the *alternation rate*, of moving the copying head to the other parent at the same index. Thus alternation copies sections of code from each parent into the child genome. In order to allow alternation to an index not identical to the prior index, we perturb the index with Gaussian noise, using the *alignment deviation*

Table 6.1 Genetic operator parameter settings used in all of our PushGP runs

Parameter	Value
Uniform mutation rate	0.01
Constant tweak rate	0.5
Uniform close mutation rate	0.1
Close increment rate	0.2
Alternation rate	0.01
Alignment deviation	10

as the standard deviation for the perturbation. Thus the copy head may jump forward or backward during an alternation, but will not likely jump far.

Finally, we employ genetic operator pipelines to chain together two or more operators to create a single child. We mainly use this functionality to create a child genome by applying alternation and then uniform mutation.

The genetic operators described here meet the requirements of uniformity described in Sect. 6.1. In particular, all three operators give uniform probability of inheriting particular genetic material in a parent: with uniform mutation and uniform close mutation this probability is explicitly defined and with alternation it is roughly one half. Additionally, alternation allows "arbitrary combinations of components from each parent to appear in the child" [18], even though some of these combinations may be more likely than others based on position in the parent.

6.5 Automatic Code Blocks Experiment

One of the primary motivations for developing Plush was to ensure that control instructions, which take arguments from the `exec` stack, have code blocks as arguments instead of single instructions. As discussed above, Plush automatically opens one or more parenthesized code block following each instruction that requires one or more arguments from the `exec` stack. Here, we conduct an experiment to examine the utility of automatic code blocks.

For this experiment we created a system that does not automatically create code blocks following specific instructions, but otherwise has similar characteristics to Plush. We started with Plush and removed the automatic opening of code blocks following specific instructions. We then added to the instruction set copies of `noop_open_paren`, as described in Sect. 6.3.3. Since we expect code blocks should be more common than other instructions, we added a number of copies of `noop_open_paren` to the instruction set to make a random program have a similar number of open parentheses as when using Plush with automatic parentheses; this resulted in around 30 copies added to about 150 other instructions, varying slightly per problem and instruction set. Otherwise, this method uses the same implementation as Plush, including close parenthesis markers. We call this method *Auto-Parens Off* for this experiment.

We compare Plush having automatic parentheses on and off using five general program synthesis benchmark problems. These problems require programs to manipulate multiple data types and use control flow structures. As such, we expect solutions to these problems will likely need to use hierarchical structure of code blocks in order to solve the problem, although solutions are possible without such structure. For more details about each problem, see their definitions in [1].

We conducted 100 runs with automatic parentheses on and 100 with them off on each problem. Table 6.2 gives the number of successful runs, i.e. runs that found

Table 6.2 Number of succesful runs out of 100

Problem	Plush	Auto-parens off
Replace space with newline	51	51
Negative to zero	45	34
X-word lines	8	0
Count odds	8	5
Digits	7	9

To be successful, a program has to perfectly pass all cases in the training set as well as an unseen test set. "Auto-Parens Off" is the version of the system where the locations of parentheses must be determined by evolution, instead of automatically

a solution program that passed all of the training cases as well as every case in an unseen test set. The only problem that showed a significant difference between the systems is X-Word Lines, where native normal Plush was significantly better than with auto parenthesis off; in fact the set of runs with them off found no solutions at all.

We examined a sample of the solution programs from each problem. With the exception of the Replace Space With Newline problem, every solution program made semantic use of code blocks; in other words, each contained a code block that was an argument to an instruction that manipulated the `exec` stack. This was true both of programs using automatic parentheses and those that did not. On the other hand, almost all of the solutions to the Replace Space With Newline problem did not make semantic use of code blocks.

While these results do not make a strong case for the importance of automatic code blocks, they do hint at its power, specifically on the X-Word Lines problem. In this problem, a program must take an input string and an integer X, and should print the string with exactly X words on each line. Interestingly, every solution to this problem had at least two layers of nested, semantically-meaningful code blocks, which we did not see in many of the solutions to other problems. It may be the case that finding the correct position for one set of parentheses does not drastically hinder evolution without automatic parentheses, but correctly nesting multiple sets of parentheses significantly increases the difficulty.

6.6 Uniform Genetic Operators Experiment

As described in Sect. 6.4, the linear genomes of Plush allowed us to implement uniform genetic operators that don't exhibit the drawbacks often associated with tree-based genetic operators. Here we explore the efficacy of each of these operators by comparing sets of runs using different combinations and probabilities of operators. We tested five different treatments consisting of different probabilities

Table 6.3 The probabilities of using each genetic operator to create a child for the five different treatments used in our genetic operator experiments

Treatment	Description	Alt	Uniform	Close	Alt/uniform
REG	Regular operators	0.2	0.2	0.1	0.5
NCM	No close mut.	0.22	0.22	0	0.56
NUM	No uni. mut.	0.9	0	0.1	0
NA	No alt.	0	0.9	0.1	0
OUM	Onlt uni. mut.	0	1.0	0	0

The operators, listed by their abreviations, are: "Alt" = alternation, "Uniform" = uniform mutation, "Close" = close mutation, and "Alt/Uniform" = alternation followed by uniform mutation

Table 6.4 Number of succesful runs out of 100 trials for different genetic operator treatments on five program synthesis problems; see Table 6.3 for treatment details

Problem	REG	NCM	NUM	NA	OUM
Replace space with newline	51	50	24	55	41
Syllables	18	20	7	9	7
Negative to zero	45	41	11	46	40
X-word lines	8	12	0	1	1
Count odds	8	5	0	6	1

To be successful, a program has to perfectly pass all cases in the training set as well as an unseen test set

of each genetic operator; these treatments are detailed in Table 6.3. Note that while no prior work has formally compared different settings of these operators, previous studies using Plush genomes [1, 3, 10] used the treatment REG.

We conducted trials with 100 GP runs using each treatment on five general program synthesis benchmark problems; again, see [1] for details of these problems. We present the results of these tests in Table 6.4.

While all treatments lead to relatively similar results, the treatment that performed most differently from the others is NUM (No Uniform Mutation). This treatment primarily uses alternation, with a small percentage of uniform close mutation operators. NUM had much lower success rates on all problems compared to REG, with a chi-squared significance test (with Holm correction) indicating significant differences at the 0.05 level on Replace Space With Newline, Negative To Zero, and X-Word Lines. This result indicates that uniform mutation has the largest role in determining success of PushGP on these problems.

Since runs without uniform mutation performed worst of our three treatments each leaving out a single operator, we decided to try a treatment only using uniform mutation, with results in the OUM column of Table 6.4; note that this treatment is very similar to NA in operator probabilities. While the success rates are lower than REG across the board, they are not significantly worse than REG on any problem. Even so, these results indicate that uniform mutation is not sufficient to produce the best results on its own, but works best in tandem with the other operators. Note that if a particular instruction disappears from the population, alternation and close mutation are not able to reintroduce it; we hypothesize that uniform mutation may provide the important ability to never get stuck in a population that cannot recover a useful lost instruction. Additionally, uniform mutation allows evolution to perform local search by changing small numbers of instructions, the importance of which has recently been noted [22].

The results do not indicate strongly whether uniform close mutation is particularly helpful or harmful. NCM, which does not use close mutation, gave results almost identical to REG. Another interesting comparison is NA, which uses 90% uniform mutation and 10% close mutation, and OUM, which uses 100% uniform mutation. While the differences between NA and OUM are not significant, NA does better on 4 of the 5 problems. Here, close mutation may be helping change the hierarchical shape of programs, which is possible through alternation but not by uniform mutation alone. Note that we never use close mutation more than 10% of the time, making it difficult to ascertain its importance.

Finally, even though these experiments show some differences between genetic operator treatments, those differences are overall minor and rarely statistically significant. These results show that the Plush representation is robust to major differences in genetic operator probabilities, as long as uniform mutation is included in some respect. This means practitioners need not worry about finding perfect settings for genetic operators, but instead can choose any reasonable settings and expect to not be worse than another setting.

6.6.1 Bloat

Code bloat without corresponding improvement in fitness has long caused problems in genetic programming [8, 17]. In our experience with uniform genetic operators in Plush, we have not observed code bloat. Figure 6.1, for example, plots the mean program size of each population over time for each run using the REG genetic operators. One of the problems (Count Odds) shows a slight *decrease* in average program size, and one (Replace Space With Newline) shows moderate growth. The mean program size for the other three problems remains fairly flat. In these runs, the maximum program size limit for the Count Odds and Negative to Zero problems is 1000, and for the other three problems is 1600. None of the mean program sizes come close to approaching these limits.

Of the uniform genetic operators we describe, only alternation can create a child of a different size than its parent. In fact, alternation has a slight bias toward creating smaller children than their parents, since alternation terminates upon reaching the end of the current parent's genome. This bias may partially account for the bloat control observed here, though children of alternation may also be larger than their parents. On the other hand, even operators that do not change the size of the produced children such as uniform mutation may have anti-bloat effects. For example, a bloated program will likely have more of its instructions replaced by

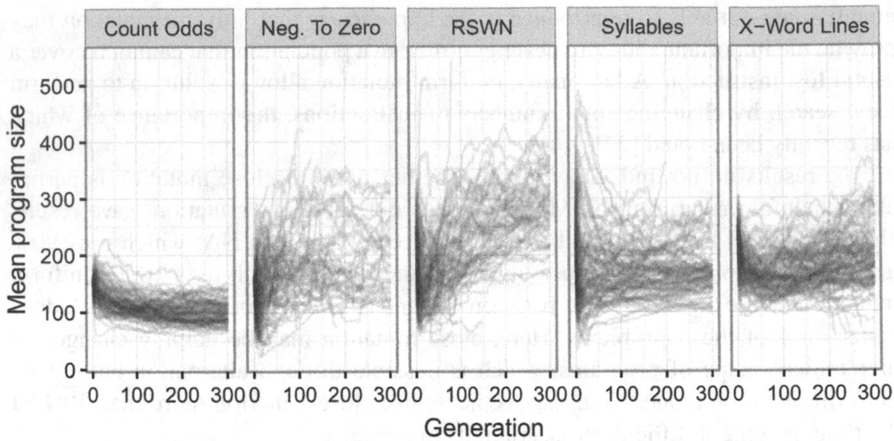

Fig. 6.1 Mean program sizes each generation for 100 runs each of five different software synthesis benchmark problems. Each run is plotted as a distinct line. "RSWN" is an abbreviation for "Replace Space With Newline"

uniform mutation than a smaller program, increasing the chances of changes that break the functionality of the parent.

6.7 Conclusions

We have described a linear representation (Plush) for structured programs (in the Push programming language), and shown that it allows for uniform genetic operators that produce meaningful structure while solving difficult problems. The central idea of the representation scheme is to use epigenetic markers, attached to instructions and literals, to indicate where structure should be added to programs when they are expressed from the linear genomes. We compared the efficacy of different combinations of uniform genetic operators operating on Plush genomes and showed how the Plush-to-Push translation scheme encourages the expression of programs with structure in appropriate places. We note that the Plush-based system appears to be relatively robust to settings of the genetic operators and that it is capable of solving difficult software synthesis problems without producing significant code bloat.

Acknowledgements This material is based upon work supported by the National Science Foundation under Grants No. 1129139 and 1331283. Any opinions, findings, and conclusions or recommendations expressed in this publication are those of the authors and do not necessarily reflect the views of the National Science Foundation.

References

1. Helmuth, T., Spector, L.: General program synthesis benchmark suite. In: GECCO '15: Proceedings of the 2015 on Genetic and Evolutionary Computation Conference, pp. 1039–1046. ACM, Madrid (2015). http://doi.acm.org/10.1145/2739480.2754769
2. Helmuth, T., Spector, L., Martin, B.: Size-based tournaments for node selection. In: Nicolau, M. (ed.) GECCO 2011 Graduate Students Workshop, pp. 799–802. ACM, Dublin (2011). https://doi.org/10.1145/2001858.2002095
3. Helmuth, T., McPhee, N.F., Spector, L.: Lexicase selection for program synthesis: a diversity analysis. In: Genetic Programming Theory and Practice XIII, Genetic and Evolutionary Computation. Springer, Berlin (2015)
4. Koza, J.R.: Genetic Programming: On the Programming of Computers by Means of Natural Selection. MIT Press, Cambridge, MA (1992). http://mitpress.mit.edu/books/genetic-programming
5. La Cava, W., Spector, L.: Inheritable epigenetics in genetic programming. In: Riolo, R., Worzel, W.P., Kotanchek, M. (eds.) Genetic Programming Theory and Practice XII, Genetic and Evolutionary Computation, pp. 37–51. Springer, Ann Arbor (2014)
6. La Cava, W., Spector, L., Danai, K., Lackner, M.: Evolving differential equations with developmental linear genetic programming and epigenetic hill climbing. In: GECCO Comp '14: Proceedings of the 2014 Conference Companion on Genetic and Evolutionary Computation Companion, pp. 141–142. ACM, Vancouver, BC (2014). https://doi.org/10.1145/2598394.2598491
7. La Cava, W., Helmuth, T., Spector, L., Danai, K.: Genetic programming with epigenetic local search. In: GECCO '15: Proceedings of the 2015 on Genetic and Evolutionary Computation Conference, Madrid, 2015, pp. 1055–1062
8. Luke, S.: Issues in scaling genetic programming: breeding strategies, tree generation, and code bloat. Ph.D. thesis, Department of Computer Science, University of Maryland, A. V. Williams Building, University of Maryland, College Park, MD (2000). http://www.cs.gmu.edu/~sean/papers/thesis2p.pdf
9. Luke, S., Panait, L.: A comparison of bloat control methods for genetic programming. Evol. Comput. **14**(3), 309–344 (2006)
10. McPhee, N.F., Donatucci, D., Helmuth, T.: Using graph databases to explore the dynamics of genetic programming runs. In: Genetic Programming Theory and Practice XIII, Genetic and Evolutionary Computation. Springer, Berlin (2015)
11. Oltean, M., Grosan, C., Diosan, L., Mihaila, C.: Genetic programming with linear representation: a survey. Int. J. Artif. Intell. Tools **18**(2), 197–238 (2009). https://doi.org/10.1142/S0218213009000111
12. Page, J., Poli, R., Langdon, W.B.: Smooth uniform crossover with smooth point mutation in genetic programming: a preliminary study. Technical Report CSRP 98-20, University of Birmingham, School of Computer Science (1998). ftp://ftp.cs.bham.ac.uk/pub/tech-reports/1998/CSRP-98-20.ps.gz
13. Perkis, T.: Stack-based genetic programming. In: Proceedings of the 1994 IEEE World Congress on Computational Intelligence, vol. 1, pp. 148–153. IEEE Press, Orlando, FL (1994). https://doi.org/10.1109/ICEC.1994.350025. http://citeseer.ist.psu.edu/432690.html
14. Poli, R., Langdon, W.B.: On the search properties of different crossover operators in genetic programming. In: Koza, J.R., Banzhaf, W., Chellapilla, K., Deb, K., Dorigo, M., Fogel, D.B., Garzon, M.H., Goldberg, D.E., Iba, H., Riolo R. (eds.) Genetic Programming 1998: Proceedings of the Third Annual Conference, pp. 293–301. Morgan Kaufmann, University of Wisconsin, Madison, WI (1998). http://www.cs.essex.ac.uk/staff/poli/papers/Poli-GP1998.pdf
15. Poli, R., Page, J.: Solving high-order Boolean parity problems with smooth uniform crossover, sub-machine code GP and demes. Genet. Program. Evolvable Mach. **1**(1/2), 37–56 (2000). https://doi.org/doi:10.1023/A:1010068314282. http://citeseer.ist.psu.edu/335584.html

16. Ryan, C., Collins, J.J., O'Neill, M.: Grammatical evolution: evolving programs for an arbitrary language. In: Banzhaf, W., Poli R., Schoenauer, M., Fogarty T.C. (eds.) Proceedings of the First European Workshop on Genetic Programming, Lecture Notes in Computer Science, vol. 1391, pp. 83–96. Springer, Paris (1998). https://doi.org/10.1007/BFb0055930. http://www.lania.mx/~ccoello/eurogp98.ps.gz
17. Silva, S., Costa, E.: Dynamic limits for bloat control in genetic programming and a review of past and current bloat theories. Genet. Program. Evolvable Mach. **10**(2), 141–179 (2009). https://doi.org/10.1007/s10710-008-9075-9
18. Spector, L., Helmuth, T.: Uniform linear transformation with repair and alternation in genetic programming. In: Riolo, R., Moore, J.H., Kotanchek M. (eds.) Genetic Programming Theory and Practice XI, Genetic and Evolutionary Computation, chap. 8, pp. 137–153. Springer, Ann Arbor (2013). https://doi.org/10.1007/978-1-4939-0375-7_8
19. Spector, L., Robinson, A.: Genetic programming and autoconstructive evolution with the push programming language. Genet. Program. Evolvable Mach. **3**(1), 7–40 (2002). https://doi.org/10.1023/A:1014538503543. http://hampshire.edu/lspector/pubs/push-gpem-final.pdf
20. Spector, L., Moore, R., Robinson, A.: Virtual quidditch: A challenge problem for automatically programmed software agents. In: Goodman, E.D. (ed.) 2001 Genetic and Evolutionary Computation Conference Late Breaking Papers, San Francisco, CA, 2001, pp. 384–389. http://hampshire.edu/lspector/pubs/quidditch-cite.pdf
21. Spector, L., Klein, J., Perry, C., Feinstein, M.: Emergence of collective behavior in evolving populations of flying agents. Genet. Program. Evolvable Mach. **6**(1), 111–125 (2005). https://doi.org/10.1007/s10710-005-7620-3. http://hampshire.edu/lspector/pubs/emergence-collective-GPEM.pdf
22. Trujillo, L., Z-Flores, E., Juárez-Smith, P., Legrand, P., Silva, S., Castelli, M., Vanneschi, L., Schütze, O., Muñoz, L.: Local search is underused in genetic programming. In: Genetic Programming Theory and Practice XIV, Genetic and Evolutionary Computation. Springer, Berlin (2016)
23. Van Belle, T., Ackley, D.H.: Uniform subtree mutation. In: Foster, J.A., Lutton, E., Miller, J., Ryan, C., Tettamanzi, A.G.B. (eds.) Genetic Programming, Proceedings of the 5th European Conference, EuroGP 2002. Lecture Notes in Computer Science, vol. 2278, pp. 152–161. Springer, Kinsale (2002)

Chapter 7
Neutrality, Robustness, and Evolvability in Genetic Programming

Ting Hu and Wolfgang Banzhaf

Abstract Redundant mapping from genotype to phenotype is common in evolutionary algorithms, especially in genetic programming (GP). Such a redundancy leads to neutrality, a situation where mutations to a genotype may not alter its phenotypic outcome. The effects of neutrality can be better understood by quantitatively analyzing its two observed properties, robustness and evolvability. In this chapter, we summarize our previous work on this topic in examining a compact Linear GP algorithm. Due to the choice of this particular system we can characterize its entire genotype, phenotype, and fitness networks, and quantitatively measure robustness and evolvability at the genotypic, phenotypic, and fitness levels. We then investigate the relationship between robustness and evolvability at those different levels. Technically, we use an ensemble of random walkers and hill climbers to study how robustness and evolvability are related to the structure of genotypic, phenotypic, and fitness networks and influence the evolutionary search process.

Keywords Genetic programming · Linear GP · Neutral networks · Robustness · Evolvability · Genomic diversity · Structural diversity · Behavioral diversity

7.1 Introduction

In evolutionary algorithms in general, but especially in genetic programming (GP), a redundant genotype-to-phenotype mapping is common where multiple unique genotypes map to the same phenotype [1, 12, 16, 23, 25]. A related notion of

---

T. Hu (✉)
Department of Computer Science, Memorial University, St. John's, NL, Canada
e-mail: ting.hu@mun.ca

W. Banzhaf
Department of Computer Science, Memorial University, St. John's, NL, Canada
BEACON Center of the Study of Evolution in Action, Michigan State University, East Lansing, MI, USA
e-mail: banzhaf@mun.ca

R. Riolo et al. (eds.), *Genetic Programming Theory and Practice XIV*, Genetic and Evolutionary Computation,
https://doi.org/10.1007/978-3-319-97088-2_7

neutrality has been put forward to describe the mutational connectivity amongst those genotypes mapped to the same phenotype [2]. Specifically, if a single-point mutation changes one genotype to another without altering the phenotypic outcome, the mutation is called neutral. Redundancy and neutrality are different but closely related concepts. Redundancy is needed for, but does not guarantee, neutrality. It is possible that although a phenotype can be represented by multiple genotypes, these genotypes cannot be traversed from one to the other through single-point mutations. In such a case, any mutations applied to those genotypes will change their phenotype.

Neutrality appears as an embedded property of many evolutionary algorithms and its influence on evolution has seen many debates in the field of evolutionary computation. On the one hand, neutrality may seem to hamper the evolutionary search since neutral mutations are not phenotypically effective [7, 27]. On the other hand, neutrality is considered beneficial for the search by providing a buffer against deleterious mutations [32] and, more importantly, by offering mutational potential through expanding neutral genotypic regions which are not subject to selection pressure [14, 29]. These two aspects relate neutrality to two notions that have drawn much attention in studies on both computational and natural evolution, namely to robustness and evolvability.

Robustness describes the resilience of an evolutionary system in the face of constant genetic and environmental perturbations, while *evolvability* captures the ability for generating novel and adaptive phenotypes. These two properties may seem contradictory at first glance, but are commonly observed coexisting in living organisms, and are both results of neutrality.

The interplay between robustness and evolvability has been a focus of research in evolutionary biology. Both theoretical [20, 30] and empirical studies [10, 19, 21] have been put forth to elucidate the relationship between them. Using RNA molecules, some argued that neutral mutational connections constrain evolution since evolution yields phenotypes which are genotypically abundant even when they are not the most fit [8]. Others argued that robustness could facilitate evolvability, and long-term innovation could only emerge in the presence of the mutational robustness [6, 9].

The relationship between robustness and evolvability is system-dependent, and it is crucially influenced by the distribution of genotypic redundancy and the mutational interconnections among phenotypes [17, 25]. Robustness promotes evolvability only if genotypic redundancy leads to more connections to different phenotypes.

A quantitative understanding of the relationship between robustness and evolvability can help resolve conflicting reports and clarify outstanding research questions. *Genotype networks*, a.k.a., neutral networks, provide a general framework for quantitatively characterizing robustness and evolvability, and have found applications in a wide array of systems [6, 11, 22, 24, 26].

In genotype networks (Fig. 7.1), vertices represent unique genotypes and mutational connections are represented as edges between pairs of genotypes. A genotype network is comprised of all genotypes that encode the same phenotype. Mutations

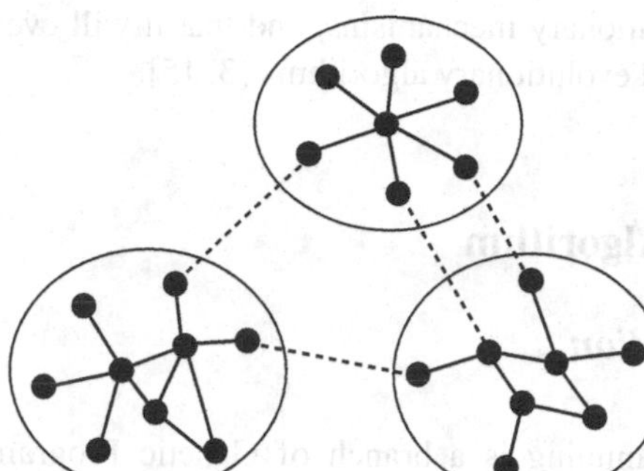

Fig. 7.1 Schematic diagram of genotype networks. Each vertex represents a genotype and all genotypes encoding the same phenotype define one genotype network. An edge links two vertices if the two genotypes can be transferred from one to another through a single point mutation. Single point mutations can also connect genotypes from different phenotypes, shown in dashed lines

within a genotype network are neutral by definition. Multiple genotype networks representing different phenotypes can also be connected through non-neutral single-point mutations. Genotype networks quantitatively characterize the distribution of genotypic redundancy among phenotypes, i.e., over-represented phenotypes have larger genotype networks and under-represented phenotypes have smaller networks. Genotype networks also capture the mutational potential among different phenotypes using different edges representing non-neutral mutations between genotypes that belong to two phenotypes.

Phenotype and fitness networks can be constructed in a similar way by representing phenotypes (or fitness values) as vertices and their mutational connections as edges. By building networks at these different levels, we are enabled to take a close look at the relationship of robustness and evolvability at the genotypic, phenotypic, and fitness levels.

Most existing studies of neutrality in evolutionary algorithms look at the effect of neutrality on the evolutionary search indirectly, i.e., they ask whether neutrality by a redundant representation improves or hampers the search ability of an evolutionary algorithm. Very little has been done to quantitatively measure robustness and evolvability directly and to study their relationship and influence on evolution dynamics.

In this chapter, we discuss the use of genotype networks to quantitatively analyze robustness and evolvability in a Linear Genetic Programming system. Linear GP has a compact representation and is especially amenable to an exhaustive enumeration of all possible genotypes and phenotypes. We characterize its genotype, phenotype, and fitness networks and their properties, and examine the diffusion and dynamics of an evolutionary population on those networks. We report on a quantitative examination of neutrality and elucidate the relationship of robustness and evolvability in GP. We hope that our analysis can find application in other GP instances and in other evolutionary algorithms, that it provides a better understanding of evolutionary mechanisms, and that it will eventually inspire new and more sophisticated evolutionary algorithms [3, 15].

7.2 Linear GP Algorithm

7.2.1 Representation

Linear Genetic Programming is a branch of Genetic Programming (GP) where the chromosomal representation is a set of instructions that are executed sequentially [5]. Although LGP follows a linear instructional structure, it is very powerful and capable of modeling complex nonlinear relationships among multiple attributes. LGP has gained increasing popularity due to its fast speed of program execution and individual evaluation [4, 13, 18, 28].

Here, we consider a two-input one-output Boolean function (Boolean circuit) modeling problem. Each LGP instruction is comprised of one return value, two operands, and one Boolean operator producing the return value from the operands. Registers $\mathtt{R}_1$ and $\mathtt{R}_2$ store the two Boolean input values. Register $\mathtt{R}_0$ takes a default initial Boolean value and its final value after the execution of all instructions is returned as the LGP program's output. To enhance the computational capacity of LGP programs, we add an extra calculation register $\mathtt{R}_3$. Calculation registers $\mathtt{R}_0$ and $\mathtt{R}_3$ can serve as either return values or as operands, whereas input registers $\mathtt{R}_1$ and $\mathtt{R}_2$ are read-only and can only serve as operands with their input content being protected from overwriting. The Boolean operator in each LGP instruction is chosen from a pre-defined operator set $opr = \{\mathtt{AND}, \mathtt{OR}, \mathtt{NAND}, \mathtt{NOR}\}$. An example Boolean LGP program with a length $L = 4$ can be given as:

$$\mathtt{R}_3 = \mathtt{R}_2 \ \mathtt{AND} \ \mathtt{R}_3$$

$$\mathtt{R}_0 = \mathtt{R}_1 \ \mathtt{OR} \ \mathtt{R}_2$$

$$\mathtt{R}_3 = \mathtt{R}_2 \ \mathtt{NAND} \ \mathtt{R}_1$$

$$\mathtt{R}_0 = \mathtt{R}_3 \ \mathtt{NAND} \ \mathtt{R}_0$$

7.2.2 Genotype, Phenotype, and Fitness

In our LGP system, the *genotype* is an unique LGP program. To enable exhaustive enumeration of the entire genotype space, we set a fixed length of $L = 4$ for all LGP programs. The total number of possible instructions is $2 \times 4 \times 4 \times 4 = 2^7$, and thus, the total number of possible four-instruction programs, i.e. genotypes, is $(2^7)^4 = 2^{28} > 268$ million. We see that even for a small problem instance and a short fixed program length the genotype space can be quite large.

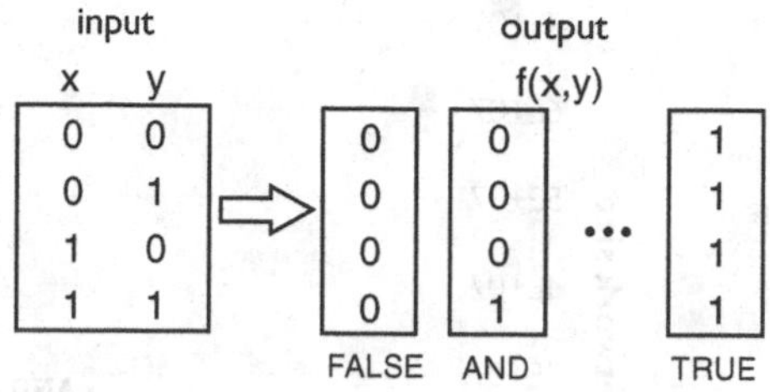

Fig. 7.2 The phenotype of a LGP program is defined as the Boolean relationship it encodes, represented as the four-digit output of the ordered two-variable inputs

We define the two-input, one-output Boolean function $f : \mathbf{B}^2 \rightarrow \mathbf{B}$, where $\mathbf{B} = \{$`TRUE`, `FALSE`$\}$ represented by a LGP program as its *phenotype*. The total number of possible phenotypes is thus $2^{2^2} = 16$. A phenotype can be represented by a set of outputs observed across each of the 2^2 possible combinations of Boolean inputs (Fig. 7.2). Compared to the large genotype space, the phenotype space is very small. This suggests a high redundancy in the mapping from genotype to phenotype, i.e., a large number of different genotypes should map to the same phenotype (approximately 16.7 million genotypes per phenotype, on average).

Based on a predefined phenotypic target, fitness can be assigned to each of the 16 phenotypes. We define the *fitness* of a phenotype to be the Hamming distance between its four-digit binary vector and that of the target. While this is technically the error between the two functions, we use the term fitness for this quantity, despite it being minimized. There are five possible fitness values, for example if the target phenotype is `TRUE` (i.e., $\langle 1111 \rangle$) the fitness of phenotype `FALSE` (i.e., $\langle 0000 \rangle$) is 4. The phenotype `x OR y` (i.e., $\langle 0111 \rangle$) has an improved fitness of 1. The mapping from phenotype to fitness is redundant again, i.e., from 16 phenotypes to five fitness values, but depends on which phenotype is set as the target. Redundancy between phenotype and fitness is less strong (approximately 3.2 phenotypes per fitness value, on average).

A single-point mutation to a genotype changes any one of the four elements of an instruction and replaces it with a randomly chosen possible allele. Single-point mutations that do not alter the phenotypic outcome are called neutral mutations. Mutations that lead to a change of phenotype are called non-neutral mutations.

7.3 Genotype, Phenotype, and Fitness Networks

Our Boolean LGP system now has 16 genotype networks, each corresponding to a particular phenotype. The distribution of genotypic redundancy is highly uneven, with the largest genotype network `FALSE` having more than 60 million genotypes (>23% of the entire genotype space) and the smallest genotype networks `x == y` and `x XOR y` having less than 25 thousand genotypes ($\ll$1% of the entire genotype space). The distribution of the sizes of genotype networks is shown in Fig. 7.3. Note that phenotype `FALSE` has more genotypes than `TRUE`, simply

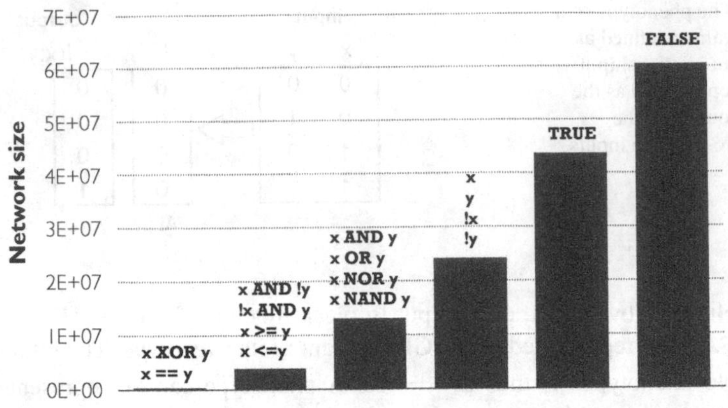

Fig. 7.3 Distribution of the size of genotype networks. Due to the symmetry of Boolean relationships, multiple phenotypes can have the same number of underlying genotypes. The size of genotype networks ranges from 25 thousand to 60 million

because all registers are initialized as FALSE before any computation. Programs whose execution does not change the content of the output register R_0 will output FALSE. The heterogeneous distribution of genotype networks suggests that some phenotypes are over-represented and some are under-represented. Random sampling and initializing genotypes likely will generate over-represented phenotypes. If a phenotypic target is under-represented, the search task will be relatively more difficult.

A phenotype network can be further constructed by representing a genotype network, i.e., a phenotype, as a vertex, and connecting two phenotypes using an edge if there exist at least one pair of genotypes of those two phenotypes that can be transferred from one to the other through a single-point mutation. Figure 7.4 shows the phenotype network in our setting. Phenotypes as vertices are numbered using the decimal values corresponding to their binary strings, labeled with their represented Boolean relationships. The phenotype network of our LGP system here is a complete graph, meaning that every vertex is connected directly to any other vertex. However, the connections are also highly heterogeneous, reflected by the varying width of edges. This suggests that a phenotype has varying mutational potentials to access other phenotypes. For instance, random mutations to genotypes of phenotype !y more likely lead to phenotypes y, FALSE, and TRUE, and less likely to phenotypes x OR y and x.

Introducing fitness further groups genotypes to build a higher-level fitness network. Since the fitness function is defined as the Hamming distance between the target phenotype and the reference phenotype, the assignment of fitness values and thus the structure of the fitness network depend on the setting of the phenotypic target. Figure 7.5 shows two fitness networks using different target phenotypes. When selection is present and rejects mutations that worsen the fitness, the fitness network becomes directed, where single-point mutations are only accepted if fitness

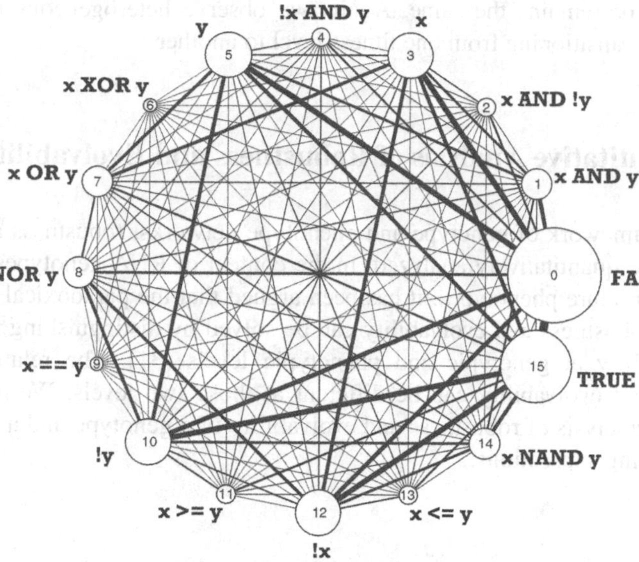

Fig. 7.4 The phenotype network. Vertices represent phenotypes and edges link two phenotypes if there exist at least one pair of genotypes mapped to the two phenotypes that can be transferred from one to the other through one single-point mutation. Vertex size is proportional to the total number of genotypes mapped to a corresponding phenotype. Edge width is proportional to the total number of one point mutations that change genotypes of one phenotype to another

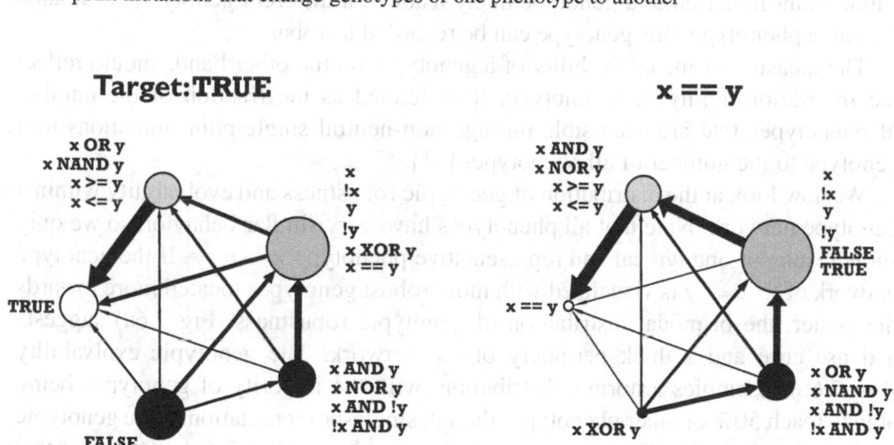

Fig. 7.5 The fitness networks with target phenotype TRUE (left) and x == y (right). Each vertex is a fitness value with white representing the best fitness zero and black representing the worst fitness four. The vertex size is again proportional to the total number of underlying genotypes. When selection prevents worsening fitness, point mutations may become irreversible, and the mutational transitions among fitness values, represented by edges, is now directed. Edge width is proportional to the total number of point mutations changing genotypes from one fitness value to another

is improved or remains the same. Again, we observe heterogeneous mutational potential for transitioning from one fitness level to another.

7.4 Quantitative Analysis of Robustness and Evolvability

Using the framework of genotype and phenotype networks, robustness and evolvability can be quantitatively analyzed. In the context of RNA genotypes and their secondary structure phenotypes, it has been argued that the paradoxical tension of mutational robustness and evolvability can be solved by distinguishing robustness and evolvability at genotypic and phenotypic levels [31]. The relationship of robustness and evolvability can be different at those two levels. We discuss the quantitative analysis of robustness and evolvability of a genotype and a phenotype in the following subsections.

7.4.1 Genotypic Robustness and Evolvability

The robustness of a genotype can be measured as the fraction of its neutral neighbors among all neighbors [31]. This definition follows the intuition that if a random single-point mutation to a genotype likely leads to a different genotype but retains the same phenotype, this genotype can be regarded as robust.

The measure of the evolvability of a genotype, on the other hand, should reflect the innovation ability of a genotype. It is defined as the fraction of the number of phenotypes that are accessible through non-neutral single-point mutations to a genotype to the number of all phenotypes [31].

We now look at the distribution of genotypic robustness and evolvability within a genotype network. Note that all phenotypes have very similar behavior, so we only show results of one typical and representative phenotype `x >= y`. If the genotype network of `x >= y` is visualized with more robust genotypes located more towards the center, the bi-modal distribution of genotypic robustness (Fig. 7.6a) suggests a dense core and a thick periphery of the network. The genotypic evolvability (Fig. 7.6b) resembles a normal distribution, with the majority of genotypes being able to reach 50% of other phenotypes though single-point mutations. The genotypic evolvability and robustness are negatively correlated (Fig. 7.6c). This negative correlation is weak ($R^2 = 0.015$) but highly significant ($p \ll 0.001$). This observation is in line with findings in RNA networks where at the genotypic level robustness and evolvability share an antagonistic relationship [31]. It is also intuitive that if random mutations to a genotype do not change its phenotype most of the time, this genotype may have less access to other different phenotypes.

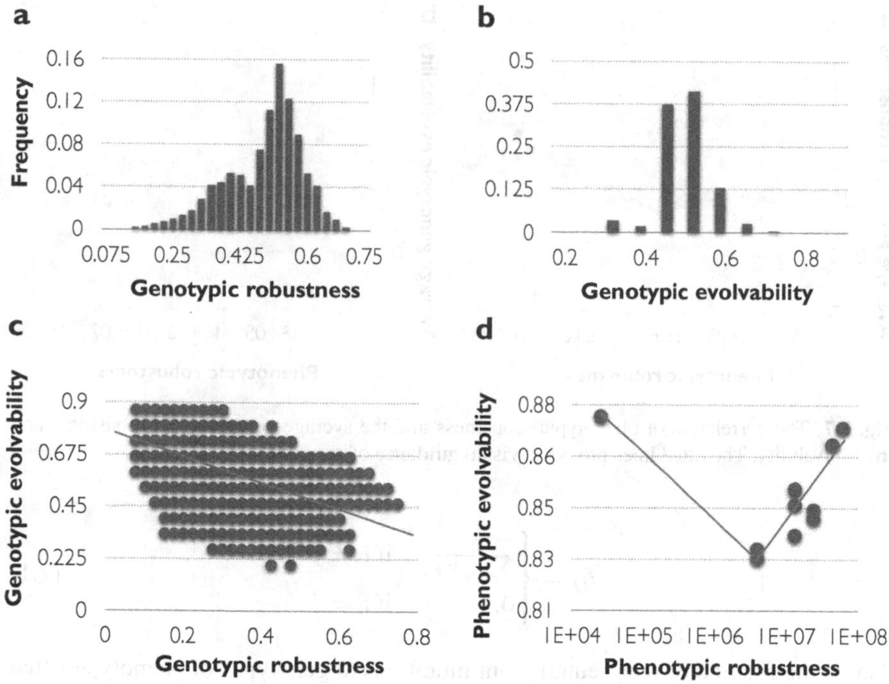

Fig. 7.6 Robustness and evolvability of genotypes and phenotypes. A typical and representative phenotype `x>=y` is chosen to show the genotypic properties in its genotype network. (**a**) and (**b**) show the distributions of genotypic robustness and evolvability, and (**c**) shows the scattered plot and correlation of genotypic robustness and evolvability. The correlation of phenotypic robustness and evolvability is shown in (**d**). The fitted lines provide a visual guidance of correlations

7.4.2 Phenotypic Robustness and Evolvability

The robustness of a phenotype is defined as the size of its genotype network, i.e., the total number of unique genotypes that map to the phenotype. The more genotypes a phenotype has, the more robust it appears.

The definition of phenotypic evolvability has seen different proposals. It can be defined similarly to genotypic evolvability as the fraction of different phenotypes that can be reached via non-neutral single-point mutations from a given phenotype [31]. However, given the complete connectivity of our phenotype network (Fig. 7.4), this phenotypic evolvability measure will assign the same value of 1 to all phenotypes.

Alternatively, the evolvability of a phenotype can be measured as the distribution of its mutational potential to other phenotypes [8]. Specifically, we use v_{ij} to denote the total number of non-neutral single-point mutations between phenotypes i and j. Letting

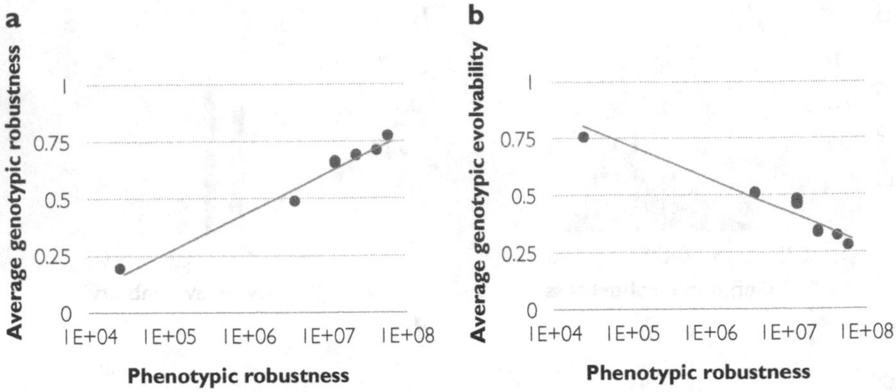

Fig. 7.7 The correlation of phenotypic robustness and the average genotypic (**a**) robustness and (**b**) evolvability. The fitted lines provide a visual guidance of correlations

$$f_{ij} = \begin{cases} \frac{v_{ij}}{\sum_{k \neq i} v_{ik}}, & \text{if } i \neq j \\ 0, & \text{if } i = j \end{cases} \tag{7.1}$$

denote the fraction of non-neutral point mutations to genotypes of phenotype i that result in genotypes of phenotype j, we define the evolvability E of a phenotype i as

$$E_i = 1 - \sum_j f_{ij}^2. \tag{7.2}$$

A phenotype that has more equally distributed mutational potential to other phenotypes is regarded as more evolvable.

The correlation of phenotypic robustness and evolvability is shown in Fig. 7.6d for all 16 phenotypes. Phenotypic robustness and evolvability are non-monotonically correlated, with median robust phenotypes having the lowest evolvability (`x AND !y` and `!x AND y`). Both the least (`x XOR y` and `x == y`) and most robust (`FALSE`) phenotypes are highly evolvable. Our results disagree with previous findings in evolutionary biology that either a monotonic positive [31] or negative [8] correlation is observed.

We also compare the measured properties across the genotypic and phenotypic levels. Figure 7.7 shows the average genotypic robustness and evolvability in relation to the phenotypic robustness. A strong and significant positive correlation ($R^2 = 0.98, p \ll 0.001$) is observed between the average genotypic robustness and the robustness of the corresponding phenotype (Fig. 7.7a). Meanwhile, average genotypic evolvability is negatively correlated ($R^2 = 0.95, p \ll 0.001$) with phenotypic robustness (Fig. 7.7b). This suggests that more robust phenotypes are comprised of more robust and less evolvable genotypes.

Note that at the level of fitness, robustness and evolvability can be defined similarly to the definition for phenotypes. However, fitness evolvability and robustness are no longer correlated (data not shown, $R^2 = 1.8 \times 10^{-4}$, $p = 0.91$).

7.5 Random Walkers and Hill Climbers

We use an ensemble of random walkers and hill climbers to investigate how the structures of genotype, phenotype, and fitness networks influence evolutionary search. We test if the quantitative measures of robustness and evolvability provide insights into predicting the search dynamics. We perform two sets of simulations. In the first set, a genotype is allowed to randomly explore the genotypic and phenotypic spaces, i.e., as a random walker. In the second set of experiments, a specific target phenotype is chosen and a fitness value is thus assigned to each genotype. Hill climbers are only allowed to move from genotypes via non-deleterious single-point mutations.

7.5.1 Random Walks Through Genotype Networks

First we investigate how individual random walkers explore the genotypic space. We consider a representative phenotype `x >= y` and confine the random walking within its genotype network. By doing so, we enforce the selection pressure on neutrality and observe its influences on evolution. Each step corresponds to a single-point mutation. We randomly pick a genotype in the phenotype `x >= y` and record all genotypes encountered in a total of four million (an approximation of the total number of genotypes in phenotype `x >= y`) steps. Then we compute the visit frequency of each genotypic robustness value during the entire course. This distribution is shown in Fig. 7.8a.

The visit frequency follows a bi-modal distribution, similar to the distribution of genotypic robustness in the genotype network of `x >= y` (Fig. 7.6a). It is true that the more frequent a robustness value is observed in a genotype network, the more likely a random walker will encounter that robustness value. So we normalize the visit frequency by dividing it by the fraction of a robustness value observed in a genotype network, i.e., by dividing Fig. 7.8a by Fig. 7.6a. The resulting distribution is shown in Fig. 7.8b.

Now we can observe a strong positive correlation of the normalized visit frequency and the genotypic robustness. This suggests that genotypes are not visited uniformly by single-point mutations, but rather in proportion to their robustness. Genotypes of high robustness are visited more often, and genotypes of low robustness are visited less often than would be expected from a random sampling of genotypes from a phenotype.

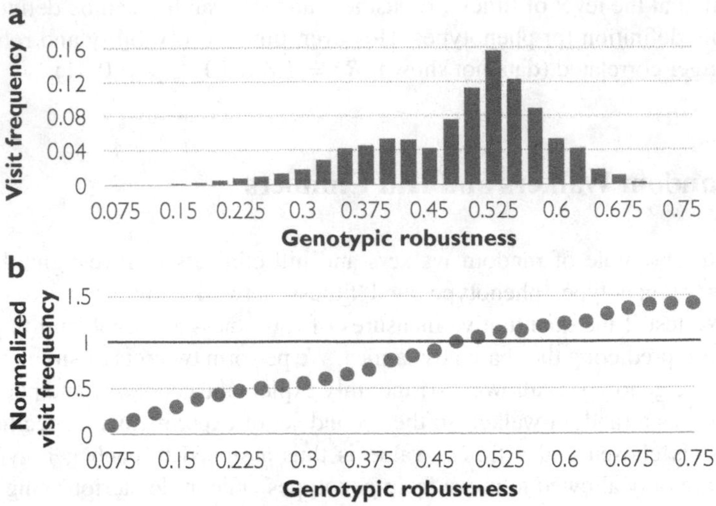

Fig. 7.8 A random walk in the genotype network of phenotype x >= y. (**a**) Distribution of the visit frequency, defined as the proportion of steps in a random walk spent at genotypes of a given robustness value. (**b**) Visit frequency normalized by the frequency with which a given genotypic robustness value is observed in the genotype network, i.e., the distribution in (**a**) divided by the distribution in Fig. 7.6a

7.5.2 Average Waiting/Adaptation Time

We set each of the 16 phenotypes as the target phenotype and one of the other 15 as the starting phenotype. For each of the 16×15 possible combinations of pairs of unique phenotypes, we then perform 1000 random walks and hill climbs, starting from a designated source phenotype and ending when the random walker or hill climber reaches any genotype in the specified target phenotype. We record the total number of point-mutations/steps required to get from one phenotype to another and calculate the average waiting (adaptation) time across 1000 random walks (hill climbs).

Figure 7.9 shows the average waiting (adaptation) time as a function of the evolvability of the source phenotype (fitness) and the robustness of the target phenotype (fitness). It is speculated that if a random walker or hill climber starts from a more evolvable phenotype, it may find a target phenotype faster. However, the evolvability of the source phenotype fails to make a prediction on the waiting time (Fig. 7.9a, $R^2 = 0.001$, $p = 0.62$), neither does the evolvability of the source fitness on the adaptation time (Fig. 7.9c, $R^2 = 0.02$, $p = 0.32$). This observation puts into question the currently available quantification of evolvability. Recall that the evolvability of a phenotype (fitness) measures the mutational potential to reach other phenotypes (fitnesses). It only captures the very first step leaving a phenotype

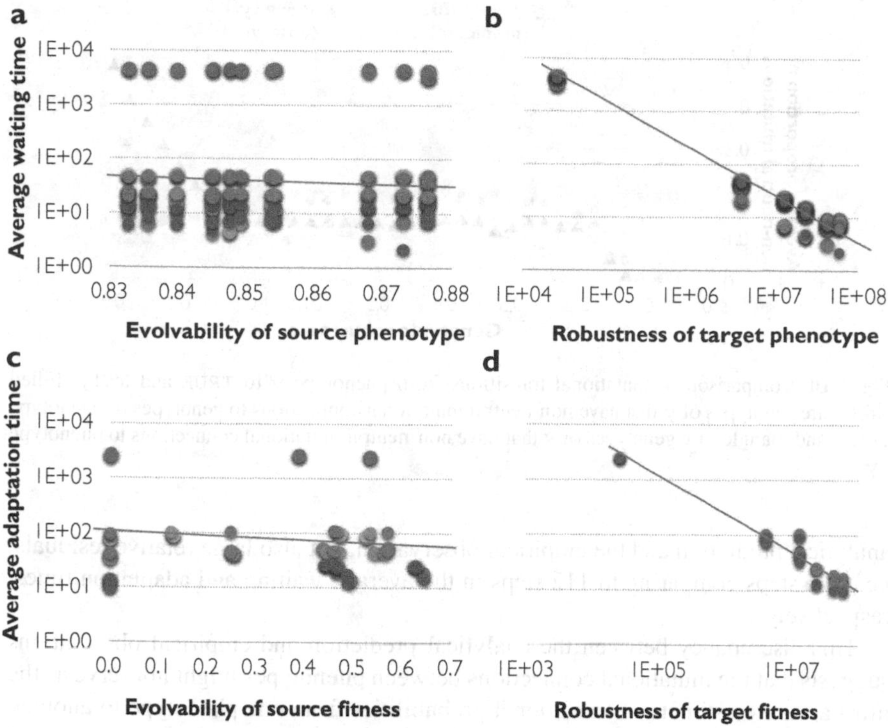

Fig. 7.9 Average waiting (adaptation) time as a function of the evolvability of the starting phenotype (fitness) and the robustness of the target phenotype (fitness)

(fitness), but fails to provide further insights on the long-term trajectory of the evolutionary process.

The robustness of the target phenotype (fitness), on the other hand, shows strong predictive power for the average waiting (adaptation) time. In Fig. 7.9b, d, a strong and negative correlation is observed between average waiting time and robustness of the target phenotype ($R^2 = 0.95$, $p \ll 0.001$), as well as between average adaptation time and robustness of the target fitness ($R^2 = 0.88$, $p \ll 0.001$). These results are intuitive and suggest that more robust phenotypes (fitnesses) are easier to reach from other phenotypes via single-point mutations since they are over-represented by more genotypes, and random mutations will more likely lead to more robust phenotypes.

The probabilistic nature of random walks and hill climbing can be captured by Markov Chain analysis, meaning that the average waiting (adaptation) time could be predicted analytically rather than through empirical simulations. By considering each phenotype as a state, and the mutational connections between phenotype i and phenotype j (f_{ij} in Eq. (7.1)) as their transition probability, we can apply Markov Chain analysis to determine the expected waiting (adaptation) time for moving from one phenotype to another. We find a strong correlation between the

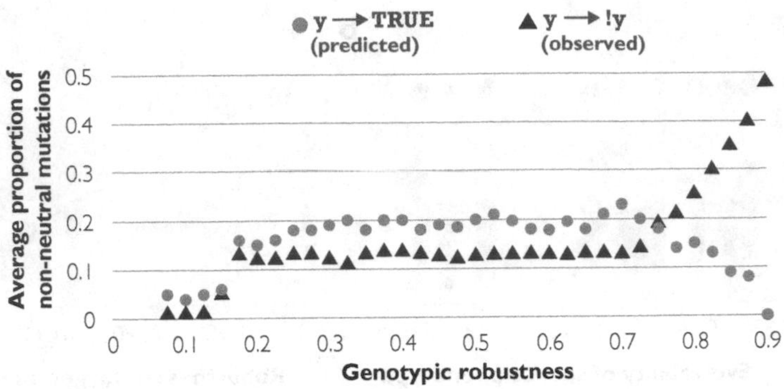

Fig. 7.10 Comparison of mutational transitions from phenotype `y` to `TRUE` and to `!y`. Filled circles are genotypes of `y` that have non-neutral mutational connections to genotypes of phenotype `TRUE`, and triangles are genotypes of `y` that have non-neutral mutational connections to phenotype `!y`

analytical prediction and the empirical observation, yet also large relative residuals, i.e., 126 steps comparing to 112 steps in the average waiting and adaptation times, respectively.

This discrepancy between the analytical prediction and empirical observations suggests that the mutational connections between phenotypes might not serve as the most accurate estimate of transitional probabilities from one phenotype to another. Let us take a close look at an example for moving from source phenotype `y` to target phenotype `!x AND y`: The predicted most likely path for this transition is `y` → `TRUE` → `!x AND y`, but the observed most frequent path is `y` → `!y` → `TRUE` → `!x AND y`, despite the fact that the observed path is longer than the predicted path and `y` has more mutational connections to `TRUE` than `!x AND y` (i.e., $f_{\mathtt{y},\mathtt{TRUE}} = 0.19$ and $f_{\mathtt{y},\mathtt{!y}} = 0.18$)!

The transitional probabilities are measured at the phenotypic level, but mutations occur at the genotypic level. Therefore, if the mutational connections between phenotypes do not provide the most accurate estimation of the transition likelihoods, a mutational bias must be introduced at the genotypic level. We take phenotypes `y`, `TRUE`, and `!y` as examples and look into the genotypes that allow a transfer from `y` to `TRUE` and to `!y`. Figure 7.10 shows the comparison of the transitions between `y` and `TRUE` and between `y` and `!y`. The non-neutral mutations connect `y` to `TRUE` (filled circles) through more genotypes with low robustness but less genotypes with high robustness, whereas the non-neutral mutations connect `y` to `!y` (filled triangles) through less genotypes with low robustness but more genotypes with high robustness. Recall that more robust genotypes are visited more frequently (Fig. 7.8b). This is the source of the bias required and explains why mutations to genotypes of `y` more likely lead to phenotype `!y` than to `TRUE`, despite the fact that the total amount of non-neutral mutations between `y` and `TRUE` is greater than that between `y` and `!y`.

7.6 Conclusion

Neutrality is commonly observed in evolutionary algorithms where mutations may not alter the phenotypic outcome. Neutrality is a result of the redundant genotype-to-phenotype mapping and debates have raged on whether neutrality is beneficial for the search ability of an evolutionary algorithm.

The effects of neutrality on its two observed properties, robustness and evolvability, can be studied quantitatively. Both robustness and evolvability capture how an evolutionary system responds to genetic changes. Robustness refers to the resilience to retain phenotypic traits in face of mutational perturbations, whereas evolvability characterizes the capability of using random mutations to generate novel and adaptive phenotypes. The relationship of robustness and evolvability may seem antagonistic, but is in fact highly collaborative.

Studying the relationship of robustness and evolvability helps to better understand the fundamental mechanisms of evolution. The framework of genotype networks has been used to quantitatively measure robustness and evolvability and to analyze their relationship. Moreover, the relationship should be better studied at the genotypic, phenotypic, and fitness levels since robustness and evolvability can take different qualifications and correlate differently at those levels.

In this book chapter, we reported on the quantitative analysis of robustness and evolvability at the genotypic, phenotypic, and fitness levels. A small-scale Linear GP system was adopted as our test system, which provides multiple advantages for our purposes. The Linear GP algorithm has a compact presentation which allows exhaustive enumeration of all possible genotypes and phenotypes. Thus the entire genotype and phenotype spaces can be characterized.

We followed evolutionary biological studies on robustness and evolvability in RNA networks and defined quantitative measures of robustness and evolvability at the genotypic, phenotypic, and fitness levels. We showed that robustness and evolvability correlate differently at those levels. At the genotypic level, a more robust genotype is less evolvable. At the phenotypic level, the correlation of robustness and evolvability is non-monotonic with the least robust and the most robust phenotypes having the highest evolvability. However, no correlation was observed at the fitness level. This finding calls for more advanced fitness evaluation methods in the future that incorporate mutational connections at the genotypic and phenotypic levels rather than simply the similarity between phenotypes.

Using an ensemble of random walkers and hill climbers, we showed how the structure of genotype, phenotype, and fitness networks can influence the evolutionary search. We found that more robust phenotypes are more accessible from other phenotypes via random mutations, however starting from a more evolvable phenotype does not guarantee a more efficient search for novel phenotypes. This is due to the limitations of evolvability measures currently available and calls for further studies.

We also found that robust genotypes play a crucial role in the evolutionary search process. More robust genotypes are visited more often than would be expected in a random sampling of genotypes, i.e., random mutations are biased leading to more robust genotypes. Therefore, robust genotypes can influence the evolutionary search by guiding it to their adjacent phenotypes. This finding is of particular interest since it may inspire mechanisms of evolutionary search that utilize robust genotypes.

Acknowledgements TH is supported by the Ignite R&D fund of Research and Development Corporation of Newfoundland and Labrador and the Canadian Natural Sciences and Engineering Research Council (NSERC) Discovery grant RGPIN-04699-2016. WB acknowledges the support from the Canadian Natural Sciences and Engineering Research Council (NSERC) Discovery grant RGPIN-283304-2012.

References

1. Banzhaf, W.: Genotype-phenotype mapping and neutral variation - a case study in genetic programming. In: Davidor, Y., Schwefel, H.P., Manner, R. (eds.) Parallel Problem Solving from Nature. Lecture Notes in Computer Science, vol. 866, pp. 322–332. Springer, Berlin (1994)
2. Banzhaf, W., Leier, A.: Evolution on neutral networks in genetic programming. In: Yu, T., Riolo, R., Worzel, B. (eds.) Genetic Programming Theory and Practice III, chap. 14, pp. 207–221. Springer, Berlin (2006)
3. Banzhaf, W., Beslon, G., Christensen, S., Foster, J.A., Kepes, F., Lefort, V., Miller, J.F., Radman, M., Ramsden, J.J.: From artificial evolution to computational evolution: a research agenda. Nat. Rev. Genet. **7**, 729–735 (2006)
4. Brameier, M.F., Banzhaf, W.: A comparison of linear genetic programming and neural networks in medical data mining. IEEE Trans. Evol. Comput. **5**(1), 17–26 (2001)
5. Brameier, M.F., Banzhaf, W.: Linear Genetic Programming. Springer, Berlin (2007)
6. Ciliberti, S., Martin, O.C., Wagner, A.: Innovation and robustness in complex regulatory gene networks. Proc. Natl. Acad. Sci. **104**(34), 13591–13596 (2007)
7. Collins, M.: Finding needles in haystacks is harder with neutrality. Genet. Program. Evolvable Mach. **7**, 131–144 (2006)
8. Cowperthwaite, M.C., Economo, E.P., Harcombe, W.R., Miller, E.L., Meyers, L.A.: The ascent of the abundant: how mutational networks constrain evolution. PLoS Computat. Biol. **4**(7), e1000110 (2008)
9. Draghi, J.A., Parsons, T.L., Wagner, G.P., Plotkin, J.B.: Mutational robustness can facilitate adaptation. Nature **463**, 353–355 (2010)
10. Ferrada, E., Wagner, A.: Protein robustness promotes evolutionary innovations on large evolutionary time-scales. Proc. R. Soc. B **275**, 1595–1602 (2008)
11. Fontana, W., Schuster, P.: Continuity in evolution: on the nature of transitions. Science **280**, 1451–1455 (1998)
12. Galvan-Lopez, E., Poli, R.: An empirical investigation of how and why neutrality affects evolutionary search. In: Cattolico, M. (ed.) Proceedings of the Genetic and Evolutionary Computation Conference, pp. 1149–1156. Springer, Berlin (2006)
13. Guven, A.: Linear genetic programming for time-series modeling of daily flow rate. J. Earth Syst. Sci. **118**(2), 137–146 (2009)
14. Hu, T., Banzhaf, W.: Neutrality and variability: two sides of evolvability in linear genetic programming. In: Rothlauf, F. (ed.) Proceedings of the Genetic and Evolutionary Computation Conference, pp 963–970 (2009)
15. Hu, T., Banzhaf, W.: Evolvability and speed of evolutionary algorithms in light of recent developments in biology. J. Artif. Evol. Appl. **2010**, 568375 (2010)
16. Hu, T., Payne, J.L., Banzhaf, W., Moore, J.H.: Robustness, evolvability, and accessibility in linear genetic programming. In: Silva, S., Foster, J.A., Nicolau, M., Machado, P.,

Giacobini, M. (eds.) Proceedings of the European Conference on Genetic Programming. Lecture Notes in Computer Science, vol. 6621, pp. 13–24. Springer, Berlin (2011)
17. Hu, T., Payne, J.L., Banzhaf, W., Moore, J.H.: Evolutionary dynamics on multiple scales: A quantitative analysis of the interplay between genotype, phenotype, and fitness in linear genetic programming. Genet. Program. Evolvable Mach. **13**, 305–337 (2012)
18. Hu, T., Moore, J.H., Banzhaf, W.: The effects of recombination on phenotypic exploration and robustness in evolution. Artif. Life **20**(4), 457–470 (2014)
19. Landry, C.R., Lemos, B., Rifkin, S.A., Dickinson, W.J., Hartl, D.L.: Genetic properties influcing the evolvability of gene expression. Science **317**, 118–121 (2007)
20. Masel, J., Trotter, M.V.: Robustness and evolvability. Trends Genet. **26**, 406–414 (2010)
21. McBride, R.C., Ogbunugafor, C.B., Turner, P.E.: Robustness promotes evolvability of thermotolerance in an RNA virus. BMC Evol. Biol. **8**, 231 (2008)
22. Reidys, C., Stadler, P.F., Schuster, P.: Generic properties of combinatory maps: neutral networks of RNA secondary structures. Bull. Math. Biol. **59**(2), 339–397 (1997)
23. Reisinger, J., Miikkulainen, R.: Acquring evolvability through adaptive representation. In: Proceedings of the Genetic and Evolutionary Computation Conference, pp. 1045–1052 (2007)
24. Rodrigues, J., Wagner, A.: Genotype networks, innovation, and robustness in sulfur metabolism. BMC Syst. Biol. **5**, 39 (2011)
25. Rothlauf, F., Goldberg, D.E.: Redundant representations in evolutionary computation. Evol. Comput. **11**(4), 381–415 (2003)
26. Schuster, P., Fontana, W., Stadler, P.F., Hofacker, I.L.: From sequences to shapes and back: a case study in RNA secondary structures. Proc. R. Soc. B **255**, 279–284 (1994)
27. Smith, T., Husbands, P., O'Shea, M.: Neutral networks and evolvability with complex genotype-phenotype mapping. In: Kelemen, J., Sosik, P. (eds.) Proceedings of the European Conference on Artificial Life, Lecture Notes in Artificial Intelligence, vol. 2159, pp. 272–281. Springer, Berlin (2001)
28. Song, D., Heywood, M.I., Zincir-Heywod, A.: A linear genetic programming approch to intrusion detection. In: Proceedings of the Genetic and Evolutionary Computation Conference. Lecture Notes in Computer Science, vol. 2724. Springer, Berlin (2001)
29. Soule, T.: Resilient individuals improve evolutionary search. Artif. Life **12**, 17–34 (2006)
30. Wagner, A.: Robustness, evolvability, and neutrality. Fed. Eur. Biochem. Soc. Lett. **579**(8), 1772–1778 (2005)
31. Wagner, A.: Robustness and evolvability: a paradox resolved. Proc. R. Soc. B **275**(1630), 91–100 (2008)
32. Yu, T., Miller, J.F.: Through the interaction of neutral and adaptive mutations, evolutionary search finds a way. Artif. Life **12**, 525–551 (2006)

Chapter 8
Local Search is Underused in Genetic Programming

Leonardo Trujillo, Emigdio Z-Flores, Perla S. Juárez-Smith, Pierrick Legrand, Sara Silva, Mauro Castelli, Leonardo Vanneschi, Oliver Schütze, and Luis Muñoz

Abstract There are two important limitations of standard tree-based genetic programming (GP). First, GP tends to evolve unnecessarily large programs, what is referred to as bloat. Second, GP uses inefficient search operators that focus on modifying program syntax. The first problem has been studied extensively, with many works proposing bloat control methods. Regarding the second problem, one approach is to use alternative search operators, for instance geometric semantic operators, to improve convergence. In this work, our goal is to experimentally show that both problems can be effectively addressed by incorporating a local search optimizer as an additional search operator. Using real-world problems, we show that this rather simple strategy can improve the convergence and performance of tree-based GP, while also reducing program size. Given these results, a question arises: Why are local search strategies so uncommon in GP? A small survey of popular GP libraries suggests to us that local search is underused in GP systems. We conclude by outlining plausible answers for this question and highlighting future work.

L. Trujillo (✉) · E. Z-Flores · P. S. Juárez-Smith · L. Muñoz
Tecnológico Nacional de México/I.T. Tijuana, Tijuana, B.C., Mexico
e-mail: leonardo.trujillo@tectijuana.edu.mx

P. Legrand
Universitè de Bordeaux, Institut de Mathèmatiques de Bordeaux, UMR CNRS 5251, Bordeaux, France

CQFD Team, Inria Bordeaux Sud-Ouest, Talence, France

S. Silva
BioISI Biosystems and Integrative Sciences Institute, Faculty of Sciences, University of Lisbon, Lisbon, Portugal

M. Castelli · L. Vanneschi
NOVA IMS, Universidade Nova de Lisboa, Lisbon, Portugal

O. Schütze
Computer Science Department, CINVESTAV-IPN, Mexico City, Mexico

R. Riolo et al. (eds.), *Genetic Programming Theory and Practice XIV*, Genetic and Evolutionary Computation,
https://doi.org/10.1007/978-3-319-97088-2_8

Keywords Genetic programming · Local search · Symbolic regression · Bloat · Neuroevolution of augmenting topologies · Numerical optimization · Evolvability

8.1 Introduction

Genetic programming (GP) is one of the most competitive approaches towards automatic program induction and automatic programming in artificial intelligence, machine learning and soft computing [18]. In particular, even the earliest version of GP, proposed by Koza in the 1990s and commonly referred to as tree-based GP or standard GP[1] [17], continues to produce strong results in applied domains over 20 years later [26, 41]. However, while tree-based GP is supported by sound theoretical insights [20, 29–31], these formalisms have not allowed researchers to completely overcome some of GP's weaknesses.

In this work, we focus on two specific shortcomings of standard GP. The first drawback is bloat, the tendency of GP to evolve unnecessarily large solutions. In bloated runs the size (number of nodes) of the best solution and/or the average size of all the individuals increases even when the quality of the solutions stagnates. Bloat has been the subject of much research in GP, comprehensively surveyed in [36]. The most successful bloat control strategies have basically modified the manner in which fitness is assigned [11, 31, 34, 37, 38], focusing the search towards specific regions of solution space.

A second problem of standard GP is the nature of the search operators. Subtree crossover and mutation operate on syntax, but are blind to the effect that these changes will have on the output of the programs, what is referred to as semantics [24]. This has led researchers to use the geometric properties of semantic space [24] and define search operators that operate at the syntax level but have a known and bounded effect on semantics, what is known as Geometric Semantic GP (GSGP). While GSGP has achieved impressive results in several domains [45], it suffers from an intrinsic shortcoming that is difficult to overstate. In particular, the sizes of the evolved solutions grow exponentially with the number of generations [24]. Since program growth is not an epiphenomenon in GSGP, as it is in standard GP, it does not seem correct to call it bloat, it is just the way that the GSGP search operates. Nonetheless, this practically eliminates one of the possible advantages of GP compared to other machine learning techniques, that the evolved solutions might be amenable to human interpretation [17, 18, 26].

The goal of this work is twofold. First, we intend to experimentally show that the effect of these problems can be substantially mitigated, if not practically eliminated, by integrating a powerful local search (LS) algorithm as an additional search operator. Our work analyzes the effects of LS on several variants of GP, including standard GP, a bloat free GP algorithm called neat-GP [43], and GSGP. In all cases,

[1] We will use the terms "standard GP" and "tree-based GP" interchangeably in this work, referring to the basic GP algorithm that relies on a tree representation and subtree genetic operators.

we will show that LS has at least one, if not several, of the following consequences: improved convergence, improved performance and reduction in program size. Moreover, we will argue that the greedy LS strategy does not increase overfitting or computational cost, two common objections towards using such approaches in meta-heuristic optimization.

The second goal of this work is to pose the following question: Why are LS strategies seldom used, if at all, in GP systems? While we do not claim that no previous works have integrated a local optimizer into a GP algorithm, the fact remains that most works with GP do not do so, with most works on the subject presenting specific application papers. This is particularly notable when we consider how ubiquitous hybrid evolutionary-LS algorithms have become, what are commonly referred to as memetic algorithms [9, 21, 25]. We will attempt to give plausible answers to this question, and to highlight important future research on the subject.

This chapter proceeds as follows. Section 8.2 discusses related work and describes our proposal to apply LS in GP for symbolic regression highlighting some experimental results. Section 8.3 shows how LS combined with a bloat-free GP can substantially reduce code growth. Afterward, Sect. 8.4 discusses recent works that apply LS with GSGP, improving convergence and performance in real-world domains. Based on the previous sections, Sect. 8.5 argues that LS strategies are underused in GP search. Finally, Sect. 8.6 presents our conclusions and future perspectives.

8.2 Local Search in Genetic Programming

Many works have studied how to combine evolutionary algorithms with LS [9, 25]. The basic idea is to include an additional operator that takes an individual as an initial point and searches for its optimal neighbor. Such a strategy can help guarantee that the local region around each individual is fully exploited. These algorithms, often called memetic algorithms, have produced impressive results in a variety of domains [9, 25].

When applying a LS strategy to GP, there are basically two broad approaches to follow: (1) apply a LS on program syntax; or (2) apply it on numerical parameters of the program. Regarding the former, [1] presents an interesting recent example. The authors apply a greedy search on a randomly chosen GP node, attempting to determine the best function to use in that node among all the possibilities in the function set. To reduce computational overhead the authors apply a heuristic decision rule to decide which trees are subject to the LS, preferring smaller trees to bias the search towards smaller solutions. A study in epigenetic repercussions for Linear GP is studied in [19], where an epigenetic activation switch is applied at the genotype level, analogously to what happens in evolutionary biology. By using a hill climber, the acquired traits are co-evolved with the original genotypes at each generation. Indeed, this can be seen as individual refinement process, or a type of LS mechanism, embedded in the broader evolutionary process.

Regarding the optimization of numerical parameters within the tree, the following works are of note. In [40] gradient descent is used to numerically optimize programs for symbolic regression problems. However, the work only optimizes the value of the terminal elements (tree leaves), it does not consider parameters within internal nodes. Similarly, in [51] and [14] a LS algorithm is used to optimize the value of constant terminal elements. In [51] gradient descent is used and tested on classification problems, applying a LS process on every individual of the population. Another recent example is [14], where Resilient Backpropagation (RPROP) is used, in this case applying the LS operator to the best individual of each generation.

From these examples, an important question for memetic algorithms is to determine when to apply the LS. For instance, [51] applies it to all the population, while [14] does so only for the best solution of each generation, and [1] uses a heuristic criterion. In the case of GP for symbolic regression, this question is studied in [49], concluding that the best strategies might be to apply LS to all the individuals in the population or a subset of the best individuals. However, that work focused on synthetic benchmarks and did not consider specialized heuristics [1]. Nonetheless, [49] does show that in general, including a LS strategy improves convergence and performance, while reducing code growth.

Other works have increased the role of the local optimizer, changing the basic GP search. Fast Function Extraction (FFX) [23], for instance, poses the symbolic regression problem as the search for the best linear combination of basis functions. Thus, FFX builds linear models, and optimizes these model using a modified version of the elastic net regression technique, eliminating the evolutionary process altogether. A similar approach can be seen in the prioritized grammar enumeration (PGE) technique [48], where dynamic programming replaces the basic search operators of traditional GP, and numerical parameters are optimized using the non-linear Levenberg–Marquardt algorithm.

8.2.1 Local Search in Symbolic Regression with Standard GP

In this work we focus on symbolic regression, though we believe that some of the conclusions might apply more generally. For now, this section describes our proposal to integrate a LS operator within GP in this domain, which we originally presented in [49, 50].

For symbolic regression, the goal is to search for the symbolic expression $K^O(\boldsymbol{\theta}^O) : \mathbb{R}^p \rightarrow \mathbb{R}$ that best fits a particular training set $\mathbb{T} = \{(\mathbf{x}_1, y_1), \ldots, (\mathbf{x}_n, y_n)\}$ of n input/output pairs with $\mathbf{x}_i \in \mathbb{R}^p$ and $y_i \in \mathbb{R}$, stated

$$(K^O, \boldsymbol{\theta}^O) \leftarrow \underset{K \in \mathbb{G}; \boldsymbol{\theta} \in \mathbb{R}^m}{arg\,min} \; f(K(\mathbf{x}_i, \boldsymbol{\theta}), y_i) \; with \; i = 1, \ldots, n \,, \tag{8.1}$$

where $\mathbb{G}$ is the solution or syntactic space defined by the primitive set $\mathbb{P}$ of functions and terminals, f is the fitness function which is based on the difference between a program's output $K(\mathbf{x}_i, \boldsymbol{\theta})$ and the desired output y_i, and $\boldsymbol{\theta}$ is a particular parametrization of the symbolic expression K, assuming m real-valued parameters. This dual problem of simultaneously optimizing syntax (structure) and parametrization can be addressed following two general approaches [12, 22]. The first group is *hierarchical structure evolution* (HSE), when $\boldsymbol{\theta}$ has a strong influence on fitness, and thus a LS is required at each iteration of the global (syntactic) search as a nested process. The second group is called *simultaneous structure evolution* (SSE), when $\boldsymbol{\theta}$ has a marginal effect on fitness, in such cases a single evolutionary loop could simultaneously optimize both syntax and parameters. These are abstract categories, but it is reasonable to state that standard GP, for instance, falls in the SSE group. On the other hand, memetic algorithms, such as the GP version we proposed in [49, 50], fall in the HSE group.

8.2.2 Proposal

First, as suggested in [16], for each individual K in the population we add a small linear uppertree above the root node, such that

$$K' = \theta_2 + \theta_1(K) , \tag{8.2}$$

K' represents the new program output, while $\theta_1, \theta_2 \in \mathbb{R}$ are the first two parameters from $\boldsymbol{\theta}$. Second, for all the other nodes n_k in the tree K we add a weight coefficient $\theta_k \in \mathbb{R}$, such that each node is now defined by

$$n'_k = \theta_k n_k , \tag{8.3}$$

where n'_k is the new modified node, $k \in \{3, \ldots, r + 2\}$ and r is the size of tree K. Notice that each node has an unique parameter that can be modified to help meet the overall optimization criteria of the non-linear expression. At the beginning of the GP run each parameter is initialized by $\theta_i = 1$. During the GP syntax search, subtrees belonging to different individuals are swapped, added or removed together with their corresponding parameters, often called Lamarckian inheritance [49, 50]. We consider each tree as a non-linear expression and the local search operator must now find the best fit parameters of the model K'. The problem could be solved using a variety of techniques, but following [49, 50] we use a trust region algorithm.

Finally, it is important to consider that the LS could increase the computational cost of the search, particularly when individual trees are very large. While applying the LS strategy to all trees might produce good results [49, 50], it is preferable to reduce the amount of trees to which it is applied. Therefore, we use the heuristic proposed in [1], where the LS is applied stochastically based on a probability $p(s)$ determined by the tree size s and the average size of the population $\overline{s}$ (details in

[1, 50]). In this way, smaller trees are more likely to be optimized than larger trees, which reduces the computational cost and improves the convergence of the optimizer by keeping the parameter vectors relatively small. Hereafter, we refer to this version of GP as GP-LS.

8.2.3 *Experiments and Results*

We evaluate this proposal on a real-world symbolic regression task, the Yacht problem, which has six features and 308 input/output samples [27]. The experiments are carried out using a modified version of the Matlab GP toolbox GPLab [35]. The GP parameters are given in Table 8.1.

In what follows, we will present results based on the median performance over all runs. The fitness function used is the RMSE, and the stopping criterion is the total number of fitness function evaluations. Function evaluations are used to account for the computational cost of the trust region optimizer, which in this case is allowed to run for 100 iterations. Results are compared with a standard GP.

Figure 8.1 summarizes the main results. The convergence plots of GP and GP-LS are shown in Fig. 8.1a, showing the median training and testing performance. The figure clearly shows that GP-LS converges faster to a lower error, and at the end of the run it substantially outperforms standard GP, consistent with [49, 50]. Figure 8.1b presents a scatter plot (each point is one individual) of all individuals generated in all runs. The individuals are plotted based on function evaluations and size. Each individual is color coded using a heat map based on test performance, with the best individuals (lowest error) in dark gray. Figure 8.1b shows that the best performance is achieved by the largest individuals.

Table 8.1 GP parameters

Parameter	Value
Runs	30
Population	100
Function evaluations	2,500,000
Training set	70% of complete data
Testing set	30% of complete data
Crossover operator	Standard subtree crossover, 0.9 prob.
Mutation operator	Mutation probability per node 0.05
Tree initialization	Full, max. depth 6
Function set	$\{+, -, \times, \div, exp, sin, cos, log, sqrt, tan, tanh\}$
Terminal set	Input features, constants
Selection for reproduction	Tournament selection of size 7
Elitism	Best individual survives
Maximum tree depth	17

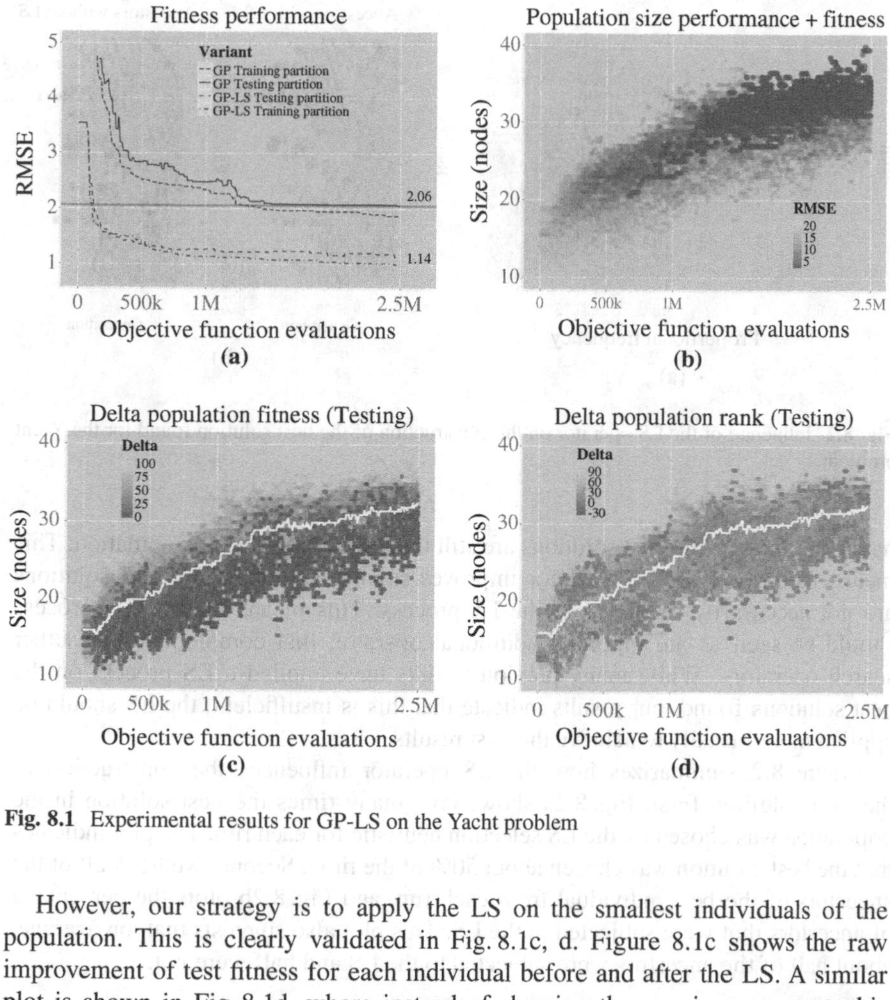

Fig. 8.1 Experimental results for GP-LS on the Yacht problem

However, our strategy is to apply the LS on the smallest individuals of the population. This is clearly validated in Fig. 8.1c, d. Figure 8.1c shows the raw improvement of test fitness for each individual before and after the LS. A similar plot is shown in Fig. 8.1d, where instead of showing the raw improvement, this figure plots the improvement in rank within the population. In both plots the average program size is plotted with a white line, and individuals that were not passed to the local optimizer have a zero value. These plots reveal that: (1) most individuals below the average size are improved by the LS; and (2) the largest improvement is exhibited by individuals that are only slightly smaller than the average program size. While the effect on program size by the LS process will be further discussed in Sect. 8.3, for now it is important to state that the median of the average program size produced by standard GP on this problem is 123.576, which is substantially higher than what is shown by GP-LS.

These results present three interesting and complimentary results. First, GP-LS clearly outperforms standard GP, in terms of convergence, solution quality and average solution size. Second, the LS is clearly improving the quality of the smallest individuals in the population, in some cases substantially. On the other

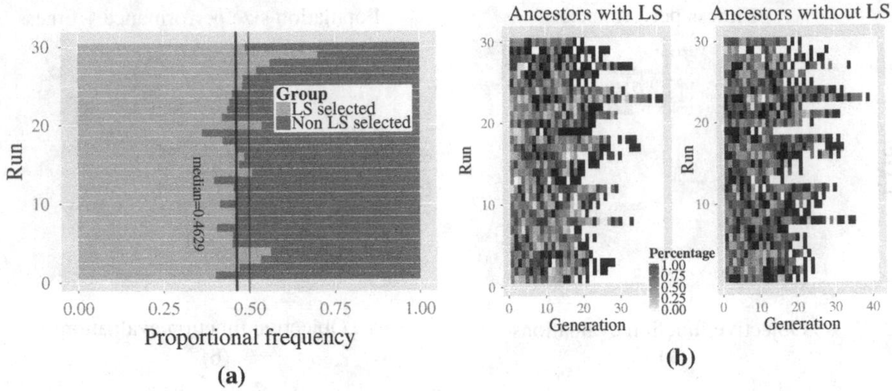

Fig. 8.2 Influence of the LS operator on the construction of the best solution found for the Yacht problem

hand, and thirdly, the best solutions are still the largest trees in the population. This means that while the LS operator improves smaller solutions, the best solutions are not necessarily subjected to the LS process. This means that the LS process should be seen as an important additional operator, that complements the other search operators. While many previous works have applied a LS process on the best solutions found, our results indicate that this is insufficient, the LS should be applied more broadly to achieve the best results.

Figure 8.2 summarizes how the LS operator influences the construction of the best solution. First, Fig. 8.2a shows how many times the best solution in the population was chosen by the LS selection heuristic for each run. The plot indicates that the best solution was chosen about 50% of the time. Second, we track all of the ancestors of the best individual from each run, and Fig. 8.2b plots the percentage of ancestors that were subjected to the LS. This plot also suggests that, on average, about half of the ancestors were subjected to the LS and half were not.

8.3 Bloat Control and Local Search

The goal of this section is to analyze the effect that the LS has on program size. We use a recently proposed bloat-free GP algorithm called neat-GP [43], which is based on the operator equalization (OE) family of methods [11, 38].

The OE approach is to control the distribution of program sizes, defining a specific shape for the distribution and then enforcing heuristic rules to fit the population to the goal distribution. Surprisingly, some of the best results are achieved by using a uniform or flat distribution; this method is called Flat-OE [37]. One of the main drawbacks of OE methods has been the difficulty of efficiently controlling the shape of the distribution without modifying the nature of the search.

Recently, neat-GP was proposed to approximate the behavior of Flat-OE in a simpler manner, exploiting well-known EC principles such as speciation, fitness sharing and elitism [43]. As the name suggests, neat-GP is designed following the general principles of the NeuroEvolution of Augmenting Topologies (NEAT) algorithm [39]. While NEAT has been used in a variety of domains, its applicability for GP in general, and for bloat control in particular, was not fully exploited until recently [42, 43].

The main features of neat-GP are the following: (a) The initial population contains trees of small depth (3 levels), the NEAT approach is to start with simple (small) solutions and to progressively build complexity (increasing size) only if the problem requires it. (b) As the search progresses, the population is divided into subsets called species, such that each species contains individuals of similar size and shape; this process is called speciation, which protects innovation during the search. (c) The algorithm uses fitness sharing, such that individuals from very large species are penalized more than individuals that belong to smaller species. This allows the search to maintain an heterogeneous population of individuals based on their size, following Flat-OE. The only exceptions are the best individuals in each species; these are not penalized allowing the search to maintain the best solutions. (d) Crossover operations mostly take place between individuals from the same species, such that the offspring will have a very similar size and shape to their parents. For a full description of neat-GP the reader is referred to [43].

8.3.1 Experiments and Results

The proposal made in this work is straightforward: combine neat-GP with the GP-LS strategy. Hereafter this hybrid method will be referred to as neat-GP-LS. The experimental work centers around comparing four GP variants: standard GP, neat-GP, GP-LS and neat-GP-LS. Each variant was applied to the real-world problems summarized in Table 8.2, which specifies the number of input variables, the size of each dataset and a general description. In this case, ten runs of each algorithm were performed, using the parameters specified in Table 8.3, using standard sub-tree mutation and crossover. For neat-GP the parameters were set according to [43]. For all algorithms, fitness and performance are computed using the RMSE. The algorithms are implemented using the DEAP library for Python [10], our code is available for download.[2]

Several changes were made to the GP-LS approach used in the previous section. First, the LS operator is applied randomly with a 0.5 probability to every individual in the population, based on the results shown in Fig. 8.2. Second, the LS operator was only allowed to run for 40 iterations, to reduce the total computational cost. Third, the termination criterion was set to 100,000 function evaluations.

[2]http://www.tree-lab.org/index.php/resources-2/downloads/open-source-tools/item/145-neat-gp.

Table 8.2 Real world regression problems used to compare all algorithms

Problem	Features	Samples	Description
Housing [33]	14	506	Housing values in suburbs of Boston
Energy cooling load [44]	8	768	Energy analysis using different building shapes simulated in Eco-tect

Table 8.3 Parameters used in all experiments

Parameter	GP Std and GP-LS	Neat-GP
Runs	30	30
Population	100	100
Function evaluations	100,000	100,000
Training set	70%	70%
Testing set	30%	30%
Crossover probability (p_c)	0.9	0.7
Mutation probability (p_m)	0.1	0.3
Tree initialization	Ramped Half-and-Half, with 6 levels of maximum depth	Full initialization, with 3 levels of maximum depth
Function set	$\{+, -, \times, sin, cos, log, sqrt, tan, tanh\}$	
Terminal set	Input variables for each real world problem, ERC	
Selection for reproduction	Tournament selection of size 7	Eliminate the worst individuals of each species
Elitism	Best individual survives	Don't penalize the best individual of each species
Maximum tree depth	17	17
Survival threshold	–	0.5
Species threshold value	–	$h = 0.15$ with $\alpha = 0.5$
LS Optimizer probability (P_s)	0.5	0.5

The algorithms are compared based on the following performance criteria: best training fitness, test fitness of the best solution, average size (number of nodes) of all individuals in the population, and size of the best solution. In particular we plot and present in table form the median performance. These results are summarized using convergence plots (performance relative to function evaluations).

Figure 8.3 summarizes the results of the tested techniques on both problems, showing convergence plots for training and testing fitness, and the average program size, each one with respect to the number of fitness function evaluations, showing the median over 10 independent runs. Notice that in this case, performance is more or less equal for all algorithms. This might be due to the different GP implementations used (GPLab and DEAP) or to the different parametrizations of the LS strategy. Nevertheless, when we inspect program size we clearly see the benefits of using the LS strategy. First, GP evolves substantially larger solutions for both problems, about one order of magnitude of difference relative to the other methods. Second,

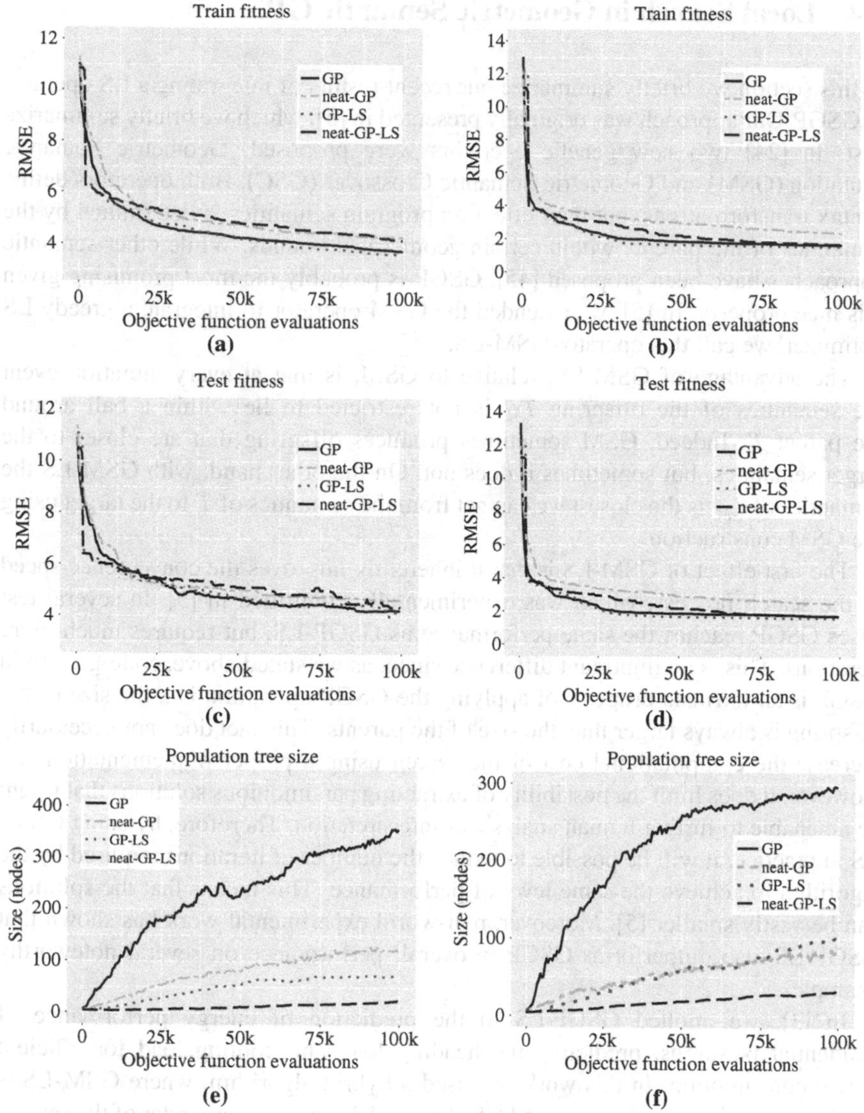

Fig. 8.3 Results for real world problems Housing **(a)**, **(c)**, **(e)** and Energy Cooling Load **(b)**, **(d)**, **(f)** plotted with respect to total function evaluations: **(a)**, **(b)** Fitness over train data; **(c)**, **(d)** Fitness over test data; and **(e)**, **(f)** Population size. Plots show median values over ten independent runs

surprisingly GP-LS is able to control code growth just as well as neat-GP. In other words, GP-LS has the same parsimonious effect on evolution as an algorithm explicitly designed for bloat control. Finally, when we combine both algorithms in neat-GP-LS, the reduction in solution size is drastically improved.

8.4 Local Search in Geometric Semantic GP

In this section we briefly summarize our recent results of integrating a LS operator to GSGP. Our approach was originally presented in [5], which we briefly summarize first. In [24] two new genetic operators were proposed, Geometric Semantic Mutation (GSM) and Geometric Semantic Crossover (GSC). Both operators define syntax transformations, but their effect on program semantics is determined by the semantics of the parents within certain geometrical bounds. While other semantic approaches have been proposed [45], GSGP is probably the most promising given this nice property. In [5], we extended the GSM operator to integrate a greedy LS optimizer, we call this operator GSM-LS.

The advantage of GSM-LS, relative to GSM, is that at every mutation event the semantics of the offspring T_M is not restricted to lie within a ball around the parent T. Indeed, GSM sometimes produces offspring that are closer to the target semantics, but sometimes it does not. On the other hand, with GSM-LS the semantics of T_M is the closest we can get from the semantics of T to the target using the GSM construction.

The first effect of GSM-LS is that it inherently improves the convergence speed of the search process, which was experimentally confirmed in [5]. In several test cases GSGP reaches the same performance as GSGP-LS, but requires much more iterations. This is an important difference since, as we stated above, code-growth in GSGP is an intrinsic property of applying the GSGP operators; i.e., the size of the offspring is always larger than the size of the parents. This fact does not necessarily increase the computational cost of the search using a proper implementation [2]. However, it does limit the possibility of extracting parsimonious solutions that might be amenable to further human analysis or interpretation. Therefore, by using GSM-LS in practice, it will be possible to reduce the number of iterations required by the algorithm to achieve the same level of performance. This means that the solutions can be vastly smaller [5]. Moreover, real-world experimental work has shown that GSGP-LS also outperforms GSGP in overall performance on several noteworthy examples.

In [3], we applied GSGP-LS to the prediction of energy performance of residential buildings, predicting the heading load and cooling load for efficient power consumption. In this work, we used a hybrid algorithm, where GSM-LS is used at the beginning of the run and GSM is used during the remainder of the search, while also performing linear scaling of the input variables. Experimental results showed that the algorithms outperformed such methods as iteratively reweighted least squares and random forests. A similar application domain was addressed in [6], where GSGP-LS was used for energy consumption forecasting. Accurate forecasting can have many benefits for electric utilities, with errors increasing costs and reducing efficiency. In this domain, GSGP-LS outperformed GSGP and standard GP, with the former considered to be a state-of-the-art method in this domain.

Then, in [4] we applied GSGP-LS to predict the relative position of computerized tomography slices, an important medical application for machine learning methods. In this work, GSGP-LS was compared with GSGP, standard GP and state-of-the-art results reported on the same problem dataset. GSGP-LS outperformed all other methods. Sometimes the difference was quite large, as much as 22% relative to other published results. We presented one final example in [7], where GSGP-LS was used to predict the per capita violent crimes in urban areas. In this case GSGP-LS was compared with linear regression, radial basis function networks, isotonic regression, neural networks and support vector machines (SVM). The only conventional algorithm that achieved equivalent performance was SVM. These examples are meant to illustrate the benefits of integrating LS into GSGP.

8.5 Discussion

Based on the results presented and discussed in Sects. 8.2–8.4, we make the following two major conclusions, limiting our discussion to the real-valued symbolic regression domain. First, integrating a numerical LS operator within a GP search brings about several benefits, including improving convergence, improving (or at least not reducing) performance, and substantially reducing code growth. We would stress the importance of the last point, the reduction in solution size is maybe the most important, if we consider the attention that bloat has received in GP literature and the potential of GP as a machine learning paradigm to generate human interpretable solutions. Moreover, program size is reduced even more when LS is combined with an explicit bloat control strategy. Second, the LS approach should be seen as an additional genetic operator, and not as a post-processing step. It seems that by subjecting the GP individuals to a numerical optimization process, the search is able to unlock the full potential of each individual. It is common to see that small individuals usually have a substantially lower fitness than larger ones, indeed this is understood as one of the reasons for bloat to appear [11]. Our results make this observation more nuanced, it is not small individuals but small individuals with sub-optimal parameterizations that will usually perform poorly. Therefore, the LS operator should be seen as a way of extracting the full potential of each GP expression, before it is kept or filtered by selection.

These conclusions seem to be supported by our experimental evidence, but we do not feel like we have hit upon an overly hidden truth or property of the GP search process. In fact, these observations seem to be relatively obvious and simple, particularly (as we have said) keeping to symbolic regression. Therefore, a question comes to mind: Why are LS strategies so uncommon in GP systems or libraries? Take for instance some of the most popular GP platforms, including for instance lilGP [32], TinyGP [28], DEAP [10], GPLab [35], ECJ [47], Open Beagle [13], Evolving Objects [15], JGAP [8] and HeuristicLab [46]. None of these software systems (to the authors' knowledge, and based on current descriptions on their respective websites) include an explicit mechanism for applying a memetic algorithmic framework as was discussed here, where a greedy numerical optimizer performs parameter tuning for GP expressions. Some of these algorithms include numerical constants, and associated mutations for these constants to search for optimal values, or post-processing functions for solution simplification and/or optimization. But even these features are quite uncommon, and are not equivalent to the type of approach we describe here.

We speculate that several different reasons are causing this, some of these reasons are practical and some are conceptual. First, it might be that integrating this functionality might be overly complex based on specific implementation details. If this is the case, we highly recommend that future versions of these or other libraries should be made amenable to numerical LS operators. Second, it might be assumed that integrating a LS operator might make the algorithm converge to local optima or make the solutions overfit the training data. While this is a valid concern, our results, in this chapter and previous publications discussed above, suggest that this does not occur. Though we admit that further evidence should be obtained, it is reasonable to assume that a GP system should at the very least allow for a LS operator to be included. Third, it may be that some consider LS to be a computational bottleneck, increasing the already long running time of GP.

While we have not fully explored this, we find that the evidence might actually point in the opposite direction. Consider that when given the same amount of fitness function calls, GP-LS seems to outperform standard GP. If we factor in the size of the evolved solutions, it is obvious that integrating a LS operator allows GP to evaluate much smaller, and by definition, more efficient GP trees (when we do not include loops). Moreover, in the case of GSGP our results suggest that no extra cost is incurred by performing the LS process [4, 5]. Finally, we feel that it may be the case that LS might not be used because it is expected that the evolutionary process should find both the solution syntax and its optimal parametrization. While the first three reasons are practical concerns, the final one is conceptual, regarding the nature of what GP is expected to do. We believe that the design of GP algorithms should exploit all of the tools at our disposal to search for optimal models in learning problems, and that greedy LS operators, particularly those from the well-established mathematical optimization community, should be considered to be an integral part of GP search.

8.6 Conclusions and Future Work

The first major conclusion to draw from this chapter—including the experimental evidence and related literature analysis—is that integrating a numerical LS operator helps to substantially improve the performance of a GP symbolic regression, based on performance, convergence, and more notably program size.

The second conclusion is that numerical LS and memetic search is seldom integrated in most GP systems. Numerical LS optimizers should be considered an important part of any GP-based search, allowing the search process to fully evaluate the usefulness of a particular GP tree before discarding it, since it very well may be that a low fitness value is due to a suboptimal parametrization of the solution.

Going forward, we identify several areas of opportunity for the GP community. The fact that memetic approaches have not been fully explored in GP literature opens up several future lines of inquiry. For instance, one line would seek to determine what is the best memetic strategy for GP, Lamarckian or Baldwinian—a choice that might be domain dependent. Another topic is to study the effect of using different LS optimization algorithms. Numerical optimizers are well suited for real-valued symbolic regression, but these methods might not generalize to other domains. Moreover, while we are assuming that the GP tree is always a non-linear expression, this may not always be the case. Therefore, other numerical optimization methods should be evaluated, based on the application domain.

The combination of both syntactic and numerical LS should also be the subject of future work, allowing us to fully exploit the local neighborhood around each solution. Moreover, while we believe that computational cost is not an issue relative to standard GP, it would still be advantageous to reduce any associated costs of the LS operator. One way, of course, is to use very efficient LS techniques or efficient implementations of these algorithms. Another possibility is to explore the development of surrogate models of LS optimization. The most important effect of the LS process is the change in relative rank for the individuals. It may be possible to derive predictive models that allow us to determine the expected effect that the LS process will have on a particular program.

Acknowledgements Funding for this work was provided by CONACYT Basic Science Research Project No. 178323 and Fronteras de la Ciencia FC-2015-2/944, TecNM (México) Research Projects 5414.14-P and 5621.15-P, and by the FP7 Marie Curie-IRSES 2013 European Commission program through project ACoBSEC with contract 439 No. 612689

References

1. Azad, R.M.A., Ryan, C.: A simple approach to lifetime learning in genetic programming based symbolic regression. Evol. Comput. **22**(2), 287–317 (2014). https://doi.org/10.1162/EVCO_a_00111. http://www.mitpressjournals.org/doi/abs/10.1162/EVCO_a_00111
2. Castelli, M., Silva, S., Vanneschi, L.: A C++ framework for geometric semantic genetic programming. Genet. Program. Evolvable Mach. **16**(1), 73–81 (2015). https://doi.org/10.1007/s10710-014-9218-0
3. Castelli, M., Trujillo, L., Vanneschi, L., Popovic, A.: Prediction of energy performance of residential buildings: a genetic programming approach. Energy Build. **102**(1), 67–74 (2015). https://doi.org/10.1016/j.enbuild.2015.05.013. http://www.sciencedirect.com/science/article/pii/S0378778815003849
4. Castelli, M., Trujillo, L., Vanneschi, L., Popovic, A.: Prediction of relative position of CT slices using a computational intelligence system. Appl. Soft Comput. (2015). https://10.1016/j.asoc.2015.09.021. http://www.sciencedirect.com/science/article/pii/S1568494615005931
5. Castelli, M., Trujillo, L., Vanneschi, L., Silva, S., Z-Flores, E., Legrand, P.: Geometric semantic genetic programming with local search. In: Silva, S., Esparcia-Alcazar, A.I., Lopez-Ibanez, M., Mostaghim, S., Timmis, J., Zarges, C., Correia, L., Soule, T., Giacobini, M., Urbanowicz, R.,

Akimoto, Y., Glasmachers, T., Fernandez de Vega, F., Hoover, A., Larranaga, P., Soto, M., Cotta, C., Pereira, F.B., Handl, J., Koutnik, J., Gaspar-Cunha, A., Trautmann, H., Mouret, J.B., Risi, S., Costa, E., Schuetze, O., Krawiec, K., Moraglio, A., Miller, J.F., Widera, P., Cagnoni, S., Merelo, J., Hart, E., Trujillo, L., Kessentini, M., Ochoa, G., Chicano, F., Doerr, C. (eds.) GECCO '15: Proceedings of the 2015 Annual Conference on Genetic and Evolutionary Computation, pp. 999–1006. ACM, Madrid (2015). http://doi.acm.org/10.1145/2739480.2754795
6. Castelli, M., Vanneschi, L., Trujillo, L.: Energy consumption forecasting using semantics based genetic programming with local search optimizer. Comput. Intell. Neurosci. **2015** (2015). https://doi.org/10.1155/2015/971908. http://www.ncbi.nlm.nih.gov/pmc/articles/PMC4464001/
7. Castelli, M., Sormani, R., Trujillo, L., Popovic, A.: Predicting per capita violent crimes in urban areas: an artificial intelligence approach. J. Ambient. Intell. Humaniz. Comput. **8**(1), 29–36 (2017). https://doi.org/10.1007/s12652-015-0334-3
8. Chen, D.Y., Chuang, T.R., Tsai, S.C.: JGAP: a java-based graph algorithms platform. Softw. Pract. Exp. **31**(7), 615–635 (2001)
9. Chen, X., Ong, Y.S., Lim, M.H., Tan, K.C.: A multi-facet survey on memetic computation. IEEE Trans. Evol. Comput. **15**(5), 591–607 (2011)
10. De Rainville, F.M., Fortin, F.A., Gardner, M.A., Parizeau, M., Gagne, C.: Deap: a python framework for evolutionary algorithms. In: Wagner, S., Affenzeller, M. (eds.) GECCO 2012 Evolutionary Computation Software Systems (EvoSoft), pp. 85–92. ACM, Philadelphia, PA (2012). https://doi.org/10.1145/2330784.2330799
11. Dignum, S., Poli, R.: Operator equalisation and bloat free GP. In: O'Neill, M., Vanneschi, L., Gustafson, S., Esparcia Alcazar, A.I., De Falco, I., Della Cioppa, A., Tarantino, E. (eds.) Proceedings of the 11th European Conference on Genetic Programming, EuroGP 2008. Lecture Notes in Computer Science, vol. 4971, pp. 110–121. Springer, Naples (2008). https://doi.org/10.1007/978-3-540-78671-9_10
12. Emmerich, M., Grötzner, M., Schütz, M.: Design of graph-based evolutionary algorithms: a case study for chemical process networks. Evol. Comput. **9**(3), 329–354 (2001)
13. Gagné, C., Parizeau, M.: Open BEAGLE: A new C++ evolutionary computation framework. In: Langdon, W.B., Cantú-Paz, E., Mathias, K., Roy, R., Davis, D., Poli, R., Balakrishnan, K., V. Honavar, G. Rudolph, J. Wegener, L. Bull, M.A. Potter, A.C. Schultz, J.F. Miller, E. Burke, N. Jonoska (eds.) GECCO 2002: Proceedings of the Genetic and Evolutionary Computation Conference, p. 888. Morgan Kaufmann Publishers, New York (2002). http://www.cs.bham.ac.uk/~wbl/biblio/gecco2002/GP272.pdf
14. Graff, M., Pena, R., Medina, A.: Wind speed forecasting using genetic programming. In: de la Fraga, L.G. (ed.) 2013 IEEE Conference on Evolutionary Computation, vol. 1, pp. 408–415. Cancun, Mexico (2013). https://doi.org/10.1109/CEC.2013.6557598
15. Keijzer, M., Merelo, J.J., Romero, G., Schoenauer, M.: Evolving objects: a general purpose evolutionary computation library. In: EA-01, Evolution Artificielle, 5th International Conference in Evolutionary Algorithms, pp. 231–244 (2001). http://www.lri.fr/~marc/EO/EO-EA01.ps.gz
16. Kommenda, M., Kronberger, G., Winkler, S., Affenzeller, M., Wagner, S.: Effects of constant optimization by nonlinear least squares minimization in symbolic regression. In: Proceedings of the 15th Annual Conference Companion on Genetic and Evolutionary Computation, pp. 1121–1128. ACM, New York (2013)
17. Koza, J.R.: Genetic Programming: On the Programming of Computers by Means of Natural Selection. MIT Press, Cambridge, MA (1992). http://mitpress.mit.edu/books/genetic-programming
18. Koza, J.R.: Human-competitive results produced by genetic programming. Genet. Program. Evolvable Mach. **11**(3/4), 251–284 (2010). https://doi.org/10.1007/s10710-010-9112-3. http://citeseerx.ist.psu.edu/viewdoc/summary?doi=10.1.1.297.6227. Tenth Anniversary Issue: Progress in Genetic Programming and Evolvable Machines
19. La Cava, W., Spector, L.: Inheritable epigenetics in genetic programming. In: Riolo, R., Worzel, W.P., Kotanchek, M. (eds.) Genetic Programming Theory and Practice XII, Genetic

and Evolutionary Computation, pp. 37–51. Springer, Ann Arbor (2014). https://doi.org/10.1007/978-3-319-16030-6_3
20. Langdon, W.B., Poli, R.: Foundations of Genetic Programming. Springer, Berlin (2002). https://doi.org/10.1007/978-3-662-04726-2. http://www.cs.ucl.ac.uk/staff/W.Langdon/FOGP/
21. Lara, A., Sanchez, G., Coello, C.A.C., Schutze, O.: Hcs: A new local search strategy for memetic multiobjective evolutionary algorithms. IEEE Trans. Evol. Comput. **14**(1), 112–132 (2010)
22. Lohmann, R.: Application of evolution strategy in parallel populations. In: International Conference on Parallel Problem Solving from Nature, pp. 198–208. Springer, Berlin (1990)
23. McConaghy, T.: Ffx: fast, scalable, deterministic symbolic regression technology. In: Riolo, R., Vladislavleva, E., Moore, J.H. (eds.) Genetic Programming Theory and Practice IX, Genetic and Evolutionary Computation, chap. 13, pp. 235–260. Springer, Ann Arbor (2011). https://doi.org/10.1007/978-1-4614-1770-5_13. http://trent.st/content/2011-GPTP-FFX-paper.pdf
24. Moraglio, A., Krawiec, K., Johnson, C.G.: Geometric semantic genetic programming. In: International Conference on Parallel Problem Solving from Nature, pp. 21–31. Springer, Berlin (2012)
25. Neri, F., Cotta, C., Moscato, P.: Handbook of Memetic Algorithms, vol. 379. Springer, Berlin (2012)
26. Olague, G., Trujillo, L.: Evolutionary-computer-assisted design of image operators that detect interest points using genetic programming. Image Vis. Comput. **29**(7), 484–498 (2011). https://doi.org/10.1016/j.imavis.2011.03.004. http://www.sciencedirect.com/science/article/B6V09-52GXV83-1/2/1462102339b445428fa4f2702939a41e
27. Ortigosa, I., Lopez, R., Garcia, J.: A neural networks approach to residuary resistance of sailing yachts prediction. In: Proceedings of the International Conference on Marine Engineering MARINE, vol. 2007, p. 250 (2007)
28. Poli, R.: TinyGP. See TinyGP GECCO 2004 competition at http://cswww.essex.ac.uk/staff/sml/gecco/TinyGP.html (2004)
29. Poli, R., McPhee, N.F.: General schema theory for genetic programming with subtree-swapping crossover: Part I. Evol. Comput. **11**(1), 53–66 (2003). https://doi.org/10.1162/106365603321829005. http://cswww.essex.ac.uk/staff/rpoli/papers/ecj2003partI.pdf
30. Poli, R., McPhee, N.F.: General schema theory for genetic programming with subtree-swapping crossover: Part II. Evol. Comput. **11**(2), 169–206 (2003). https://doi.org/10.1162/106365603766646825. http://cswww.essex.ac.uk/staff/rpoli/papers/ecj2003partII.pdf
31. Poli, R., McPhee, N.: Parsimony pressure made easy. In: Keijzer, M., Antoniol, G., Congdon, C.B., Deb, K., Doerr, B., Hansen, N., Holmes, J.H., Hornby, G.S., Howard, D., Kennedy, J., Kumar, S., Lobo, F.G., Miller, J.F., Moore, J., Neumann, F., Pelikan, M., Pollack, J., Sastry, K., Stanley, K., Stoica, A., Talbi, E.G., Wegener, I. (eds.) GECCO '08: Proceedings of the 10th Annual Conference on Genetic and Evolutionary Computation, pp. 1267–1274. ACM, Atlanta (2008). https://doi.org/10.1145/1389095.1389340. http://www.cs.bham.ac.uk/~wbl/biblio/gecco2008/docs/p1267.pdf
32. Punch, B., Zongker, D.: LiL-GP 1.1. a genetic programming system (1998)
33. Quinlan, J.R.: Combining instance-based and model-based learning. In: Proceedings of the Tenth International Conference on Machine Learning, pp. 236–243 (1993)
34. Silva, S.: Reassembling operator equalisation: a secret revealed. In: Krasnogor, N., Lanzi, P.L., Engelbrecht, A., Pelta, D., Gershenson, C., Squillero, G., Freitas, A., Ritchie, M., Preuss, M., Gagne, C., Ong, Y.S., Raidl, G., Gallager, M., Lozano, J., Coello-Coello, C., Silva, D.L., Hansen, N., Meyer-Nieberg, S., Smith, J., Eiben, G., Bernado-Mansilla, E., Browne, W., Spector, L., Yu, T., Clune, J., Hornby, G., Wong, M.L., Collet, P., Gustafson, S., Watson, J.P., Sipper, M., Poulding, S., Ochoa, G., Schoenauer, M., Witt, C., Auger A. (eds.) GECCO '11: Proceedings of the 13th Annual Conference on Genetic and Evolutionary Computation, pp. 1395–1402. ACM, Dublin (2011) https://doi.org/10.1145/2001576.2001764
35. Silva, S., Almeida, J.: Gplab-a genetic programming toolbox for matlab. In: Proceedings of the Nordic MATLAB Conference, pp. 273–278 (2003)

36. Silva, S., Costa, E.: Dynamic limits for bloat control in genetic programming and a review of past and current bloat theories. Genetic Program. Evolvable Mach. **10**(2), 141–179 (2009). https://doi.org/10.1007/s10710-008-9075-9
37. Silva, S., Vanneschi, L.: The importance of being flat-studying the program length distributions of operator equalisation. In: Riolo, R., Vladislavleva, E., Moore J.H. (eds.) Genetic Programming Theory and Practice IX, Genetic and Evolutionary Computation, chap. 12, pp. 211–233. Springer, Ann Arbor (2011). https://doi.org/10.1007/978-1-4614-1770-5_12
38. Silva, S., Dignum, S., Vanneschi, L.: Operator equalisation for bloat free genetic programming and a survey of bloat control methods. Genetic Programming and Evolvable Machines **13**(2), 197–238 (2012). https://doi.org/10.1007/s10710-011-9150-5
39. Stanley, K.O., Miikkulainen, R.: Evolving neural networks through augmenting topologies. Evol. Comput. **10**(2), 99–127 (2002)
40. Topchy, A., Punch, W.F.: Faster genetic programming based on local gradient search of numeric leaf values. In: Spector, L., Goodman, E.D., Wu, A., Langdon, W.B., Voigt, H.M., Gen, M., Sen, S., Dorigo, M., Pezeshk, S., Garzon, M.H., Burke E. (eds.) Proceedings of the Genetic and Evolutionary Computation Conference (GECCO-2001), pp. 155–162. Morgan Kaufmann, San Francisco, CA (2001). http://garage.cse.msu.edu/papers/GARAGe01-07-01.pdf
41. Trujillo, L., Legrand, P., Olague, G., Levy-Vehel, J.: Evolving estimators of the point-wise hoelder exponent with genetic programming. Inf. Sci. **209**, 61–79 (2012). https://doi.org/10.1016/j.ins.2012.04.043. http://www.sciencedirect.com/science/article/pii/S0020025512003386
42. Trujillo, L., Munoz, L., Naredo, E., Martinez, Y.: Neat, there's no bloat. In: Nicolau, M., Krawiec, K., Heywood, M.I., Castelli, M., Garcia-Sanchez, P., Merelo, J.J., Rivas Santos, V.M., Sim K. (eds.) 17th European Conference on Genetic Programming. Lecture Notes in Computer Science, vol. 8599, pp. 174–185. Springer, Granada (2014). https://doi.org/10.1007/978-3-662-44303-3_15
43. Trujillo, L., Munoz, L., Galvan-Lopez, E., Silva, S.: Neat genetic programming: controlling bloat naturally. Inf. Sci. **333**, 21–43 (2016). https://doi.org/10.1016/j.ins.2015.11.010. http://www.sciencedirect.com/science/article/pii/S0020025515008038
44. Tsanas, A., Xifara, A.: Accurate quantitative estimation of energy performance of residential buildings using statistical machine learning tools. Energy Build. **49**, 560–567 (2012)
45. Vanneschi, L., Castelli, M., Silva, S.: A survey of semantic methods in genetic programming. Genetic Program. Evolvable Mach. **15**(2), 195–214 (2014). https://doi.org/10.1007/s10710-013-9210-0. http://link.springer.com/article/10.1007/s10710-013-9210-0
46. Wagner, S., Affenzeller, M.: Heuristiclab: A generic and extensible optimization environment. In: Adaptive and Natural Computing Algorithms, pp. 538–541. Springer, Berlin, (2005)
47. White, D.R.: Software review: the ECJ toolkit. Genet. Program. Evolvable Mach. **13**(1), 65–67 (2012). https://doi.org/10.1007/s10710-011-9148-z
48. Worm, T., Chiu, K.: Prioritized grammar enumeration: symbolic regression by dynamic programming. In: Blum, C., Alba, E., Auger, A., Bacardit, J., Bongard, J., Branke, J., Bredeche, N., Brockhoff, D., Chicano, F., Dorin, A., Doursat, R., Ekart, A., Friedrich, T., Giacobini, M., Harman, M., Iba, H., Igel, C., Jansen, T., Kovacs, T., Kowaliw, T., Lopez-Ibanez, M., Lozano, J.A., Luque, G., McCall, J., Moraglio, A., Motsinger-Reif, A., Neumann, F., Ochoa, G., Olague, G., Ong, Y.S., Palmer, M.E., Pappa, G.L., Parsopoulos, K.E., Schmickl, T., Smith, S.L., Solnon, C., Stuetzle, T., Talbi, E.G., Tauritz, D., Vanneschi L. (eds.) GECCO '13: Proceeding of the Fifteenth Annual Conference on Genetic and Evolutionary Computation Conference, pp. 1021–1028. ACM, Amsterdam (2013). https://doi.org/10.1145/2463372.2463486
49. Z-Flores, E., Trujillo, L., Schuetze, O., Legrand, P.: Evaluating the effects of local search in genetic programming. In: Tantar, A.A., Tantar, E., Sun, J.Q., Zhang, W., Ding, Q., Schuetze, O., Emmerich, M., Legrand, P., Del Moral, P., Coello Coello, C.A. (eds.) EVOLVE - A Bridge between Probability, Set Oriented Numerics, and Evolutionary Computation V. Advances in Intelligent Systems and Computing, vol. 288, pp. 213–228. Springer, Peking (2014). https://doi.org/10.1007/978-3-319-07494-8_15. https://hal.inria.fr/hal-01060315

50. Z-Flores, E., Trujillo, L., Schuetze, O., Legrand, P.: A local search approach to genetic programming for binary classification. In: Silva, S., Esparcia-Alcazar, A.I., Lopez-Ibanez, M., Mostaghim, S., Timmis, J., Zarges, C., Correia, L., Soule, T., Giacobini, M., Urbanowicz, R., Akimoto, Y., Glasmachers, T., Fernandez de Vega, F., Hoover, A., Larranaga, P., Soto, M., Cotta, C., Pereira, F.B., Handl, J., Koutnik, J., Gaspar-Cunha, A., Trautmann, H., Mouret, J.B., Risi, S., Costa, E., Schuetze, O., Krawiec, K., Moraglio, A., Miller, J.F., Widera, P., Cagnoni, S., Merelo, J., Hart, E., Trujillo, L., Kessentini, M., Ochoa, G., Chicano, F., Doerr, C. (eds.) GECCO '15: Proceedings of the 2015 Annual Conference on Genetic and Evolutionary Computation, pp. 1151–1158. ACM, Madrid (2015). http://doi.acm.org/10.1145/2739480.2754797
51. Zhang, M., Smart, W.: Genetic programming with gradient descent search for multiclass object classification. In: Keijzer, M., O'Reilly, U.M., Lucas, S.M., Costa, E., Soule T. (eds.) Genetic Programming 7th European Conference, EuroGP 2004, Proceedings, Lecture Notes in Computer Science, vol. 3003, pp. 399–408. Springer, Coimbra (2004). https://doi.org/10.1007/978-3-540-24650-3_38. http://www.springerlink.com/openurl.asp?genre=article&issn=0302-9743&volume=3003&spage=399

Chapter 9
PRETSL: Distributed Probabilistic Rule Evolution for Time-Series Classification

Babak Hodjat, Hormoz Shahrzad, Risto Miikkulainen, Lawrence Murray, and Chris Holmes

Abstract The EC-Star rule-set representation is extended to allow probabilistic classifiers. This allows the distributed age-layered evolution of probabilistic rule sets. The method is tested on 20 UCI data problems, as well as a larger dataset of arterial blood pressure waveforms. Results show consistent improvement in all cases compared to binary classification rule-sets.

Keywords Genetic programming · Evolutionary computation · Probabilistic rule-sets · Distributed processing · Time-series classification · Medical diagnosis

9.1 Introduction

Rule sets utilize the notion of predicate logic and form collections of statements of the form "IF antecedent condition A is met THEN consequence B occurs". These are ideal candidate models for use in medical diagnostic applications due to their explicit, interpretable structure and their ability to uncover nonlinear relationships and interactions in large data domains. The interpretability of rules is a vital attribute for medical applications, where predictions need to be auditable so that experts can understand how and why a recommendation or forecast was made. The ability of rule sets to deal naturally with nonlinear relations and interactions is another key attraction. The recent emergence of large-scale genetic epidemiology case-control studies has taught us that simple genotype–phenotype models can only explain a small proportion of the known heritable (genetic) risk component of a disease.

B. Hodjat (✉) · H. Shahrzad · R. Miikkulainen
Sentient Technologies, San Francisco, CA, USA
e-mail: babak@sentient.ai; hormoz@sentient.ai; risto.miikkulainen@sentient.ai

L. Murray · C. Holmes
University of Oxford, Oxford, UK
e-mail: murray@stats.ox.ac.uk; cholmes@stats.ox.ac.uk

R. Riolo et al. (eds.), *Genetic Programming Theory and Practice XIV*, Genetic and Evolutionary Computation,
https://doi.org/10.1007/978-3-319-97088-2_9

Probabilistic rule sets hold great potential for uncovering cryptic relationships and can maximize the use of available information contained in the data.

Up to now, rule-set models have been hampered by the computational resources that are needed to implement them effectively. In addition, there has been no way to accommodate uncertainty into rule-set predictions, so that they cannot be statistically characterized. The computational challenge of rule sets arises from the enormous search space of potential rules that might apply for any particular system, due to all the possible combinations of antecedents and consequences. Conventional optimization methods are ill-suited to scale to such spaces.

Age-varying fitness calculation is an approach suitable for data problems in which evolved solutions need to be applied to many fitness samples in order to measure a candidate's fitness confidently [5]. This approach is elitist: best candidates of each generation are retained to be run on more fitness cases to improve confidence in the candidate's fitness. The number of fitness evaluations in this method depends on the relative fitness of a candidate solution compared to others at any given point.

EC-Star [8] is a massively distributed evolutionary platform that uses age-varying fitness as the basis for distribution. This allows for easier distribution of big-data problems through sampling or hashing/feature reduction techniques: breaking the data stash into smaller chunks, each chunk contributing to the overall evaluation of the candidates.

In this paper, the power of EC-star search is combined with a probabilistic extension of rule-based logic into a new method called PRETSL (Probabilistic Rule Evolution for Time-Series cLassification). In a probabilistic rule set the consequences of rules are used to update a conditional probability statement. For example, a probabilistic rule might be, "IF condition *A* is present THEN the probability of the disease occurring increases by *Z*", where *A* and *Z* are parameters to be learned by the system. The probabilities suggested by all rules of the set are combined and thresholded to produce the final classification.

The EC-Star platform and related work in probabilistic classifiers is first reviewed below. The PRETSL approach for using fuzzy logic and probabilistic rule-sets in an age-layered distributed evolutionary run is then outlined. Initial results are presented from experimentation on 20 data sets from the UCI collection, as well as on an application on a blood-pressure prediction task, comparing the probabilistic with a binary classifier rule-set representation. The results suggest that PRETSL is an effective approach, making it possible to combine knowledge at a more fine-grained level, and thus increasing classification accuracy.

9.2 Prior Work

In EC-Star, age is defined as the number of fitness samples upon which a candidate has been evaluated. This system uses a hub-and-spoke architecture for distribution, where the main evolutionary process is moved to the processing nodes (Fig. 9.1).

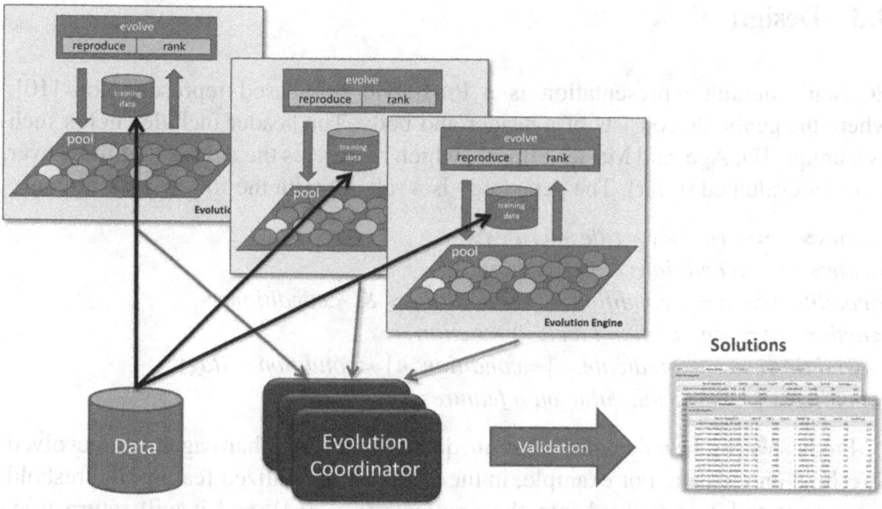

Fig. 9.1 The EC-Star hub-and-spoke distribution architecture. Each Evolution Engine runs an independent evolution on its own pool of candidates on a limited amount of data, and periodically reports the results to the Evolution Coordinator. The Evolution Coordinator maintains a list of the best candidates found so far and periodically sends them back to the Evolution Engines for further evaluation. In this manner, EC-Star utilizes age-layering to speed up evolution, and takes advantage of heterogeneous and potentially unreliable computing resources across the internet

Each node, or Evolution Engine, has its own pool of candidate rule-sets, or individuals, and independently runs through the evolutionary cycle. At each new generation, an Evolution Engine submits its fittest candidates to the Evolution Coordinator (i.e., the server) for consideration. This step takes place typically after a set number of evaluations, referred to as the maturity age.

The server side, or Evolution Coordinator, maintains a list of the best of the best candidates so far. EC-Star achieves scale through making copies of genes at the server, sending them to Evolution Engines for aging, and merging the aged results received back from them (Fig. 9.1). This process also allows the spreading of the fitter genetic material. EC-Star is massively distributable by running each Evolution Engine on a processing node (e.g., CPU) with limited bandwidth and occasional availability [6]. Typical runs utilize hundreds of thousands of processing units spanning across thousands of geographically dispersed sites. In the Evolution Coordinator, only candidates of the same age range are compared with one another (thus implementing age-layering). Each age range has a fixed quota.

EC-Star has previously been used, for example, in the blood-pressure prediction task, and was found to be an effective implementation for rule evolution on time-series data sets—a class of problems that is not as well suited for traditional classification methods such as Random Forest [3]. In this paper, it will be extended into probabilistic classification.

9.3 Design

EC-Star's default representation is a Pitts-style rule-based representation [10], where the genotype consists of a header and body. The header includes fields such as a unique ID, Age, and Master Fitness (which represents the aggregate fitness over samples evaluated so far). The gene body is a rule set with the following grammar:

<rules> ::= *<rule>* | *<rule><rules>*

<rule> ::= *<conditions>* → *<action>*

<conditions> ::= *<condition>* | *<condition>* & *<conditions>*

<action> ::= *<prediction label>* | *<action>*

<condition> ::= *<predicate>* |*<condition>* | *<condition>* [*lag*]

<predicate> ::= *<truth value on a feature>*

Predicates can be calculated as an inequality (e.g., less-than) against an evolved threshold on the data. For example, in the case of a normalized feature, a threshold between 0 and 1 is evolved into the predicate (say, 0.4), and it will return true, should the inequality (i.e., *feature* < 0.4) evaluate to true in the presence of that threshold.

EC-Star allows for applying fuzzy logic [7] to the evaluation of predicates and rules. The fuzzy value for a predicate inequality is derived by applying a sigmoid function on the inequality: The closer the feature is to the threshold, the closer the resulting continuous value is to 1. Fuzzy logic is then used to calculate a fuzzy value for the rule as a whole.

In order to represent a probabilistic rule-set [2], an action is defined to be an evolvable probability between 0 and 1, representing the likelihood of a sample to belong to a class label defined over the data-set. In its simplest form, the probabilities of different rules that fire over a data sample are aggregated into a single probability for a binary classification system. For example, if three rules fire, returning 0.2, 0.4, and 0.6 respectively, the output verdict on the sample can be calculated as the average of the probabilities (0.4). Taking the fuzzy logic value of each rule into account gives us the opportunity to calculate the rule-set verdict as a weighted average using the fuzzy value of each rule as the weight. The last step, if needed by the domain, is to convert this value to the binary classification, with 0.5 as the threshold.

Note that it is possible that, for a given fitness sample, no rules fire. In such a case—depending on the problem domain requirements—either a default action is selected, or the fitness sample is said to have resulted in a no-action state. The no-action state can thus be treated separately in the fitness function.

The fitness of a probabilistic rule-set is then calculated as the mean absolute error (MAE) of its predictions. Below, this method is referred to as PRETSL, for Probabilistic Rule Evolution for Time-Series cLassification.

9.4 Experiments

First, the PRETSL approach is demonstrated on 20 standard UCI data sets [1]. Each data set consists of a number of data points (patients), each with a number of predictors (e.g., biometrics and health history), some of which are missing for each data point. These data are partitioned randomly such that the training set has roughly 70% of data points, and the remaining 30% data points are withheld and used as the test set.

The EC-Star platform is used to train 50 binary and 50 probabilistic classifiers using the training set. For each entry in the dataset, the binary classifiers output $z = P(y = 1|x) \in \{0, 1\}$, while the probabilistic classifiers output $z = P(y = 1|x) \in [0, 1]$; in both cases, output z values are interpreted as the problem-specific predicted probability (e.g., that the patient does not survive the study).

To compare methods, the mean squared error (MSE) of each classifier's predictions is calculated using the test set:

$$\mathrm{MSE}(d^v) = \frac{1}{N^v} \sum_{n=1}^{N^v} (P(y = 1|x = x_n^v) - y_n^v)^2. \tag{9.1}$$

Note that in the case of hard classifiers, this measure reduces to the misclassification rate.

Figures 9.2, 9.3, and 9.4 give the results. The PRETSL approach improves classification performance in every single case and is comparable to results from random forest.

Second, PRETSL is demonstrated on a much larger real world problem of classifying time series of arterial blood pressure (ABP) data. Our particular area of investigation is acute hypotensive episodes.

A large number of patient records are time series based. Some are at the granularity of high resolution physiological waveforms recorded in the ICU or via the remote monitoring systems. Given a time-series of training exemplars each of length T (in samples), to build a discriminative model capable of predicting an event, features are extracted by splitting the time series into non-overlapping (or overlapping), divisions of size k samples each, up to a certain point $h < T$ such that there are $m = h/k$ divisions. A number of aggregating functions are then applied to each of these divisions (a.k.a windows) to give features for the problem.

The blood-pressure dataset consists of roughly 4000 patient's ABP waveforms from MIMIC II v3, with a sampling rate of 125 Hz [4], recorded invasively from one of the radial arteries. The raw data size is roughly one terabyte. The labels in the data are imbalanced; the total number of Low events is just 1.9% of the total number of events. In total, there are 45,693 EC-Star data packages from 4414 patient records. Of these, 32,898 packages with 100 data points each (i.e., events) were used as the training set and 12,795 samples as the test set.

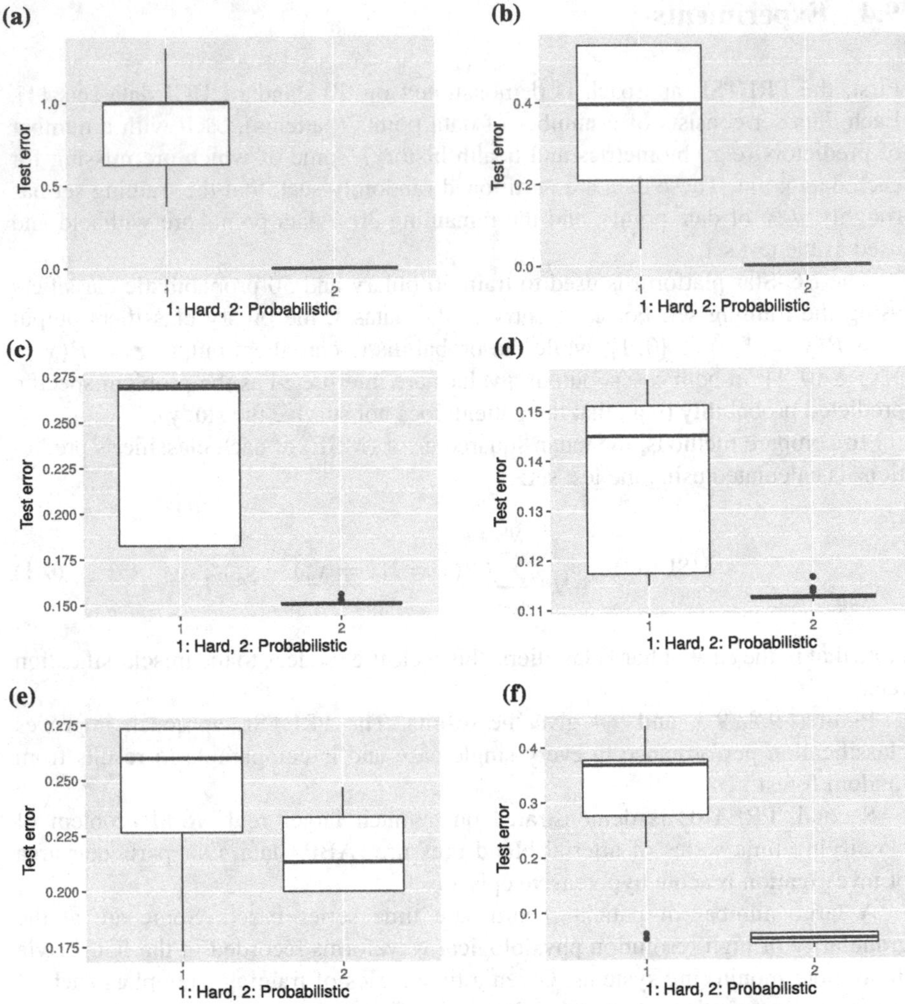

Fig. 9.2 Distribution of MSE on the test set for the 50 binary (i.e. hard) classifiers and the 50 probabilistic classifiers (i.e. PRETSL) for the first six of the 20 UCI datasets. The probabilistic classifiers outperformed the binary classifiers in each case in this figure as well as in Figs. 9.3 and 9.4, demonstrating the advantage of the PRETSL approach. (**a**) Acute inflammation. (**b**) Acute nephritis. (**c**) Adult income. (**d**) Bank marketing. (**e**) Blood transfusion. (**f**) Breast cancer Wisc Diag

Figure 9.5 gives the results, again showing that the probabilistic classifiers outperform the binary classifiers. Indeed, the worst performing soft classifier outperforms the best of the hard classifiers. An example probabilistic rule-set evolved by the system is given below, where V_n represents features from the wavelets in the data set, and prob is the probability for the patient to have developed critically low blood pressure after a 30 min blackout window:

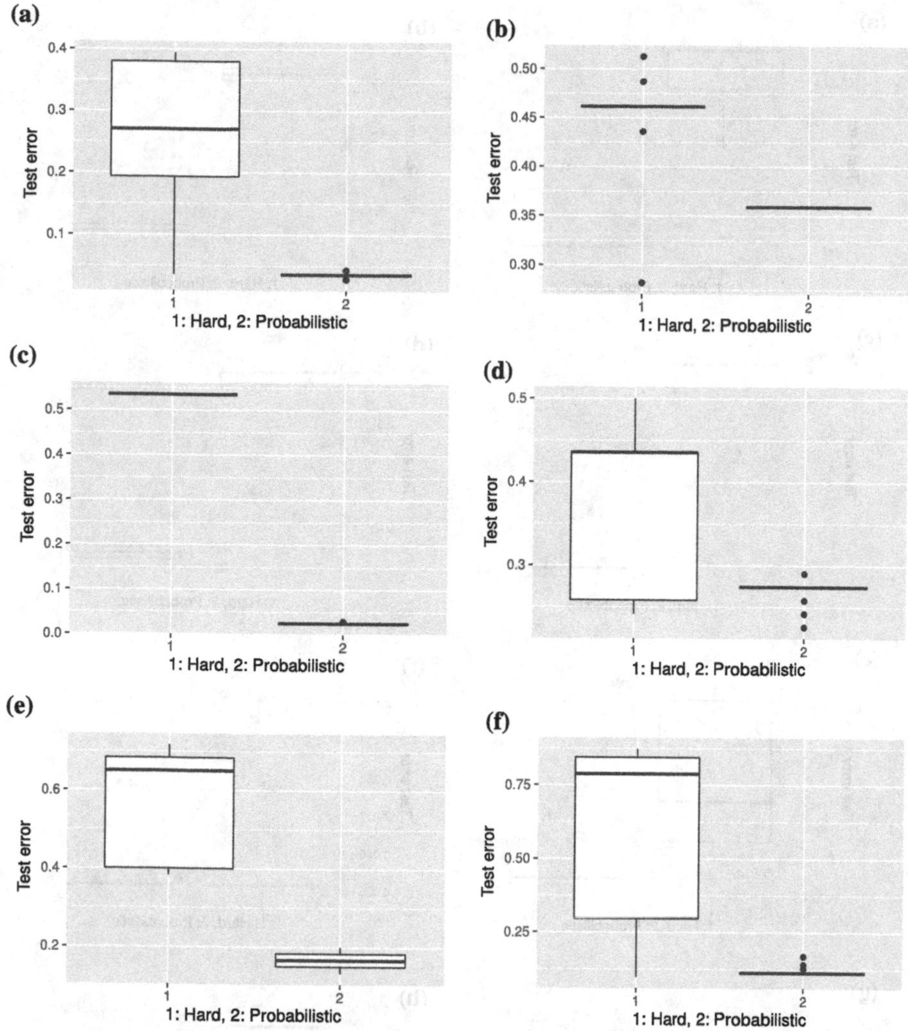

Fig. 9.3 *(continued from Fig. 9.2)* Results for the second six of the 20 UCI datasets. (**a**) Breast cancer Wisc Orig. (**b**) Breast cancer Wisc Prog. (**c**) Chess KRVKP. (**d**) Haberman survival. (**e**) Heart disease cleveland. (**f**) Ionosphere

$(!V_4 < 35.13 \wedge V_{78} < 176.75 \wedge V_{52} < 6) \Longrightarrow prob = 0.04$
$(V_{78} < 79.3 \wedge V_{36} < 3.09 \wedge V_{69} < 0.08 \wedge !V_{38} < -0.25) \Longrightarrow prob = 0.95$
$(V_{78} < 79.3 \wedge V_{36} < 3.09 \wedge !V_{38} < -2.61) \Longrightarrow prob = 0.95$
$(V_{78} < 128.03 \wedge V_{63} < 0) \Longrightarrow prob = 0.14$
$(V_1 < 143.24) \Longrightarrow prob = 0.84$

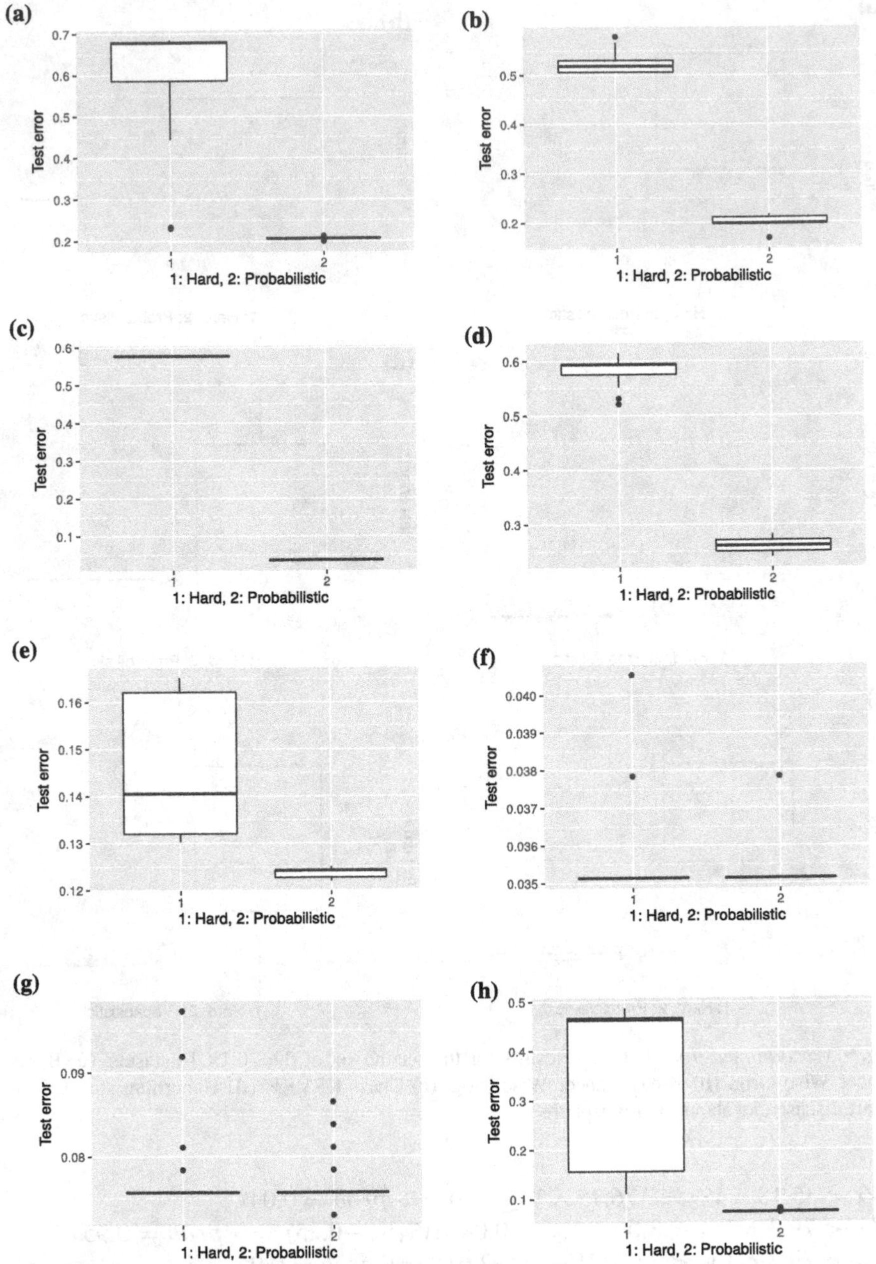

Fig. 9.4 *(continued from Fig. 9.3)* Results for the remaining eight of the 20 UCI datasets. (**a**) Magic telescope. (**b**) Mammographic masses. (**c**) Monks3. (**d**) Musk1. (**e**) Musk2. (**f**) Ozone 1 h. (**g**) Ozone 8 h. (**h**) Spambase

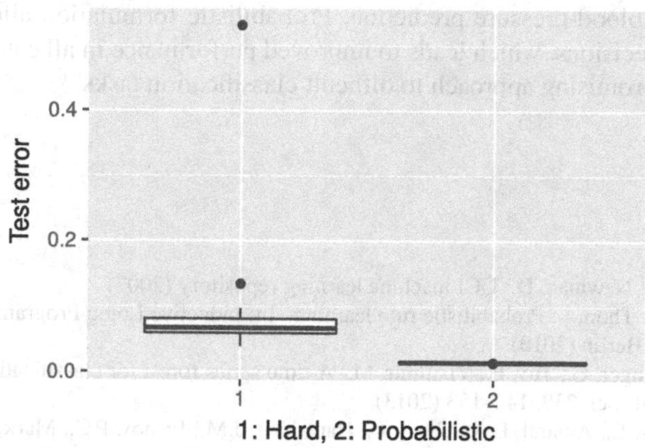

Fig. 9.5 Distribution of MSE on the test set for the 50 binary (i.e. hard) classifiers and the 50 probabilistic classifiers (i.e. PRETSL) for the MIMIC arterial blood pressure dataset. All PRETSL classifiers outperformed all binary classifiers in this scale-up experiment, demonstrating the power of the PRETSL approach in challenging problems in general, and time-series classification in particular

9.5 Discussion and Future Work

One key advantage of probabilistic predictions is that they can be combined with a formal loss function for misclassification in order to make optimal risk-based decisions, such as whether a patient should be given a new drug, or whether the patient requires further tests to make an accurate diagnosis or prognosis. Such an extension will allow for the integration of rule set models directly into the clinic.

Note that the rule sets are readily interpretable and may provide scientific insight; their probabilistic combination reduces the risk of overfitting that accompanies the use of a single classifier, and may facilitate model selection and hypothesis testing.

By framing the rule sets within a probabilistic system, formal methods from Bayesian statistics can be utilized to combine predictions across the population of rule sets in a coherent fashion [9]; this approach should improve the performance further in future work.

More work is also in order to determine the source of consistently improved performance of PRETSL versus binary classification, as demonstrated in the experiments above.

9.6 Conclusion

In this paper, evolution of rule sets for classification tasks is extended into probabilistic rule sets. This method, PRETSL, is implemented in the EC-Star distributed computing platform and evaluated in 20 UCI datasets as well as in a scale-up application of blood-pressure prediction. Probabilistic formulation allows making more refined decisions, which leads to improved performance in all cases. PRETSL is therefore a promising approach to difficult classification tasks.

References

1. Asuncion, A., Newman, D.: UCI machine learning repository (2007)
2. De Raedt, L., Thon, I.: Probabilistic rule learning. In: Inductive Logic Programming, pp. 47–58. Springer, Berlin (2010)
3. Deng, H., Runger, G., Tuv, E., Vladimir, M.: A time series forest for classification and feature extraction. Inf. Sci. **239**, 142–153 (2013)
4. Goldberger, A.L., Amaral, L.A., Glass, L., Hausdorff, J.M., Ivanov, P.C., Mark, R.G., Mietus, J.E., Moody, G.B., Peng, C.K., Stanley, H.E.: Physiobank, physiotoolkit, and physionet components of a new research resource for complex physiologic signals. Circulation **101**(23), e215–e220 (2000)
5. Hodjat, B., Shahrzad, H.: Introducing an age-varying fitness estimation function. In: Genetic Programming Theory and Practice X, pp. 59–71. Springer, Berlin (2013)
6. Hodjat, B., Hemberg, E., Shahrzad, H., OReilly, U.M.: Maintenance of a long running distributed genetic programming system for solving problems requiring big data. In: Genetic Programming Theory and Practice XI, pp. 65–83. Springer, Berlin (2014)
7. Klir, G., Yuan, B.: Fuzzy Sets and Fuzzy Logic, vol. 4. Prentice Hall, Upper Saddle River, NJ (1995)
8. OReilly, U.M., Wagy, M., Hodjat, B.: Ec-star: a massive-scale, hub and spoke, distributed genetic programming system. In: Genetic Programming Theory and Practice X, pp. 73–85. Springer, Berlin (2013)
9. Polson, N.G., Scott, J.G., Windle, J.: Bayesian inference for logistic models using pólya–gamma latent variables. J. Am. Stat. Assoc. **108**(504), 1339–1349 (2013)
10. Smith, S.F.: A learning system based on genetic adaptive algorithms (1980)

Chapter 10
Discovering Relational Structure in Program Synthesis Problems with Analogical Reasoning

Jerry Swan and Krzysztof Krawiec

Abstract Much recent progress in Genetic Programming (GP) can be ascribed to work in semantic GP, which facilitates program induction by considering program behavior on individual fitness cases. It is therefore interesting to consider whether alternative decompositions of fitness cases might also provide useful information. The one we present here is motivated by work in analogical reasoning. So-called proportional analogies ('gills are to fish as lungs are to mammals') have a hierarchical relational structure that can be captured using the formalism of Structural Information Theory. We show how proportional analogy problems can be solved with GP and, conversely, how analogical reasoning can be engaged in GP to provide for problem decomposition. The idea is to treat pairs of fitness cases as if they formed a proportional analogy problem, identify relational consistency between them, and use it to inform the search process.

Keywords Genetic programming · Program synthesis · Proportional analogy · Inductive logic programming · Machine learning

10.1 Introduction

Perhaps the strongest reason for favouring Genetic Programming (GP) over alternative machine learning approaches is the explanatory power afforded by the resulting symbolic descriptions. Whilst other approaches may be faster or more accurate, GP can provide more compelling insights into observed data than numerically-driven approaches constrained to specific model class.

J. Swan
Department of Computer Science, University of York, York, UK

K. Krawiec (✉)
Institute of Computing Science, Poznan University of Technology, Poznań, Poland
e-mail: krzysztof.krawiec@cs.put.poznan.pl

R. Riolo et al. (eds.), *Genetic Programming Theory and Practice XIV*, Genetic and Evolutionary Computation,
https://doi.org/10.1007/978-3-319-97088-2_10

To maximize the explanatory power of GP, it is highly desirable to obtain symbolic explanations which appear to the human reader to be not only comprehensible but also natural. In respect of comprehensibility, there has been considerable work in combating expression bloat [38]. However, there has been relatively little emphasis on building human bias into the search process. Since much human bias originates in universal observations that stem from the specific constitution of the natural world, its inclusion may actually lead to both quantitative and qualitative improvements [40]. Since GP is often used to search for regularities in real-world data, equipping it with such biases may be desirable, at the least in extracting more compelling explanations from experimental results [39].

In this chapter, we explore a mechanism for the discovery of problem's relational structure, framed in terms of existing work on analogical reasoning. Analogy can be considered as 'a mapping between systems or processes' and has been described as 'the core of cognition' [13]. In cognitive science, it is understood to provide a flexible mechanism for re-contextualising situations in terms of prior (or hypothetical) experience and is also considered a key mechanism for escaping dichotomies of representation [30], which is argued to be of general importance for Computational Intelligence [21].

We start with a brief overview of analogy as a computational mechanism in Sect. 10.2. In Sect. 10.3, we present the formalism of Structural Information Theory for building the relational structures needed for the proposed approach. In Sect. 10.4, we present GPCAT, a framework for solving proportional analogy problems using GP, and experimentally assess its performance in Sect. 10.5. In Sect. 10.6, we explain how similar mechanisms can be used to aid GP applied to conventional program synthesis problems. In Sect. 10.7 we discuss the related work, and summarize this study in Sect. 10.8.

10.2 Analogical Reasoning

The use of analogy as a computational mechanism dates back to Evans's famous geometric reasoner [9]. More recent computational models include the Structure Mapping Engine (SME) [10], the connectionist models ACME [14] and LISA [15], Heuristic-Driven Theory Projection [37] and some matching techniques used in Case-Based Reasoning [1]. A short article can only provide a brief overview of the wide range of literature: considerably more detail is available in the recent volume by Prade and Richard [32]. As distinct from *predictive* analogy, which is concerned with inferring properties of a target object as a function of its similarity to a source object, our interest here is in the application of *proportional* analogy.

The roots of analogical proportion can be traced as far back as Aristotle [3]. A proportional analogy problem, denoted $A : B :: C : D$, is concerned with finding D such that D is to C as B is to A. The 'microdomain' of Letter String Analogy (LSA) Problems (e.g. *abc : abd :: ijk : ?*) can be considered exemplary and is of longstanding interest: although seemingly simple, the domain can require

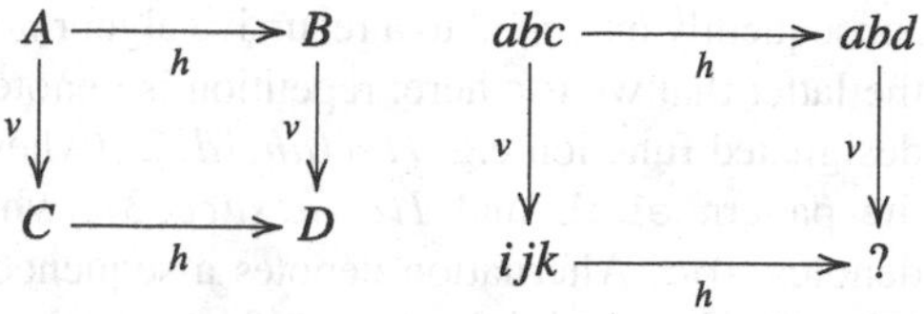

Fig. 10.1 Commutativity of proportional analogy [36]

remarkable sophistication [11]. As can be seen in Fig. 10.1, proportional analogy problems can also be considered to form a commutative diagram [36].

Notable approaches to LSA problems include Hofstadter and Mitchell's CopyCat [13] and the Anti-Unification based approach of Weller and Schmid [45]. When studied in the context of core AI research and cognitive science, LSAs are often left 'open-ended':

```
abc : abd  ::  ijk     : ?
abc : abd  ::  iijjkk  : ?
abc : abd  ::  mrrjjj  : ?
abc : abd  ::  xyz     : ?
```

Posed in this way, LSAs are unlike traditional instances of computational problem solving—in general, a LSA problem has no singular 'Platonic' solution, so it is therefore difficult to define an objective measure for solution quality in a 'top down' fashion. Nevertheless, humans confronted with LSA problems typically converge on a few answers that occur with relatively stable frequencies. For instance, the most common answers to the above LSAs are respectively `ijl`, `iijjll`, `mrrkkk` and `xya`, which corroborates the existence of human bias.

10.3 Capturing Relational Structure

Any method that is intended to deal with proportional analogy problems requires some (formal or informal) means of capturing the relational structure of objects in the domain (here: letter strings). Ideally, such a mechanism should take into account the natural biases discussed in Sect. 10.1. One means of representing and quantifying such bias is via the use of Structural Information Theory (SIT) [23]. SIT is a formalism of relational structure which also provides a complexity metric. In contrast to the complexity metrics of Algorithmic Information Theory (e.g. Kolmorogorov), SIT is explicitly designed to correspond to the principles of human Gestalt perception [8], intended to explain human propensity to prefer certain perceptual groupings. The rules of Gestalt are readily illustrated in visual perception, where they explain the inclination for grouping smaller objects into larger shapes, grouping objects by proximity, closing partially occluded curves, and others.

The original description of SIT due to Leeuwenberg [23] describes linear, one-dimensional patterns of objects in terms of repetition, alternation and symmetry, subsequently extended to a recursive algebraic description by Dastani et al. [4]. It is the latter that we use here: repetition is denoted by the iterated application of some designated function e.g. $Iter(ab, id, 2)$ (where id is the identity function) denotes the pattern abab and $Iter(a, succ, 3)$ (where $succ$ is the successor function) denotes abc. Alternation denotes a sequence into which an object is interleaved. It has 'left' and 'right' variants: for example, $Alt_L(a, xyz)$ describes axayaz and $Alt_R(a, xyz)$ describes xayaza. Symmetry denotes a sequence followed by its reversal, and occurs in an 'even' form ($Sym_E(ab)$ = abba) and an 'odd' form ($Sym_O(ab, c)$ = abcba).

A SIT term determines a unique string, but the opposite does not hold: a given sequence may clearly be representable by many different SIT descriptions. For example, the sequence abccba can be represented both by $Sym_E(Iter(a, succ, 3))$ and $Sym_O(ab, Iter(c, id, 2))$. Associated with each structural description is the notion of *information load*, intended to quantify human preference between alternative relational descriptions—those with lower information loads being preferable. The measure of information load we adopt here is due to Dastani et al. [4], which modifies the previous formulation of VanDerHelm and Leeuwenberg [43] and is defined as the sum of occurrences of individual operators in a SIT description, not including the SIT operators themselves. Thus, while $Iter(ab, id, 2)$ and $Alt_L(a, bb)$ both represent SIT descriptions of *abab*, the former has an information load of 2 and the latter 3.

10.3.1 Finding SIT Terms with GP

In the recursive variant of SIT described above, the patterns appearing in a SIT relation above can themselves be SIT relations. This lends itself to a direct representation of SIT relations as nodes in a tree structure, allowing the use of GP to find a SIT description for a given string [4]. As mentioned above, it is desirable to search for SIT structures of low complexity, as given by the information load measure. However, this quantity alone cannot effectively drive the search, as the relations found by GP have to produce the target string in the first place. Therefore, we define our fitness function as:

$$f(t) = Lev(t, s) + 0.001 \times InfLoad(t), \tag{10.1}$$

where t is the SIT term being evaluated, s is the string to be reproduced, $Lev(t, s)$ is the Levenshtein distance between the string produced by t and s, and $InfLoad(t)$ is the information load. The fitness function effectively realizes lexicographic ordering of search objectives, prioritizing matching the target string. Alternatively, a multiobjective evolutionary search could be engaged here.

The instruction set of our GP setup includes all the algebraic relations presented above, i.e., $Iter$, Alt_R, Alt_L, Sym_E, Sym_O, and the $Sequence$ and $Group$ relations that respectively cater for flat and nested (hierarchical) sequences. There are also numeric constants that $Iter$ needs to determine the number of iterations and function literals: $succ$, $pred$, and id. Terms, numeric constants and function literals form three types handled by strongly-typed GP mechanisms. Using EpochX GP [29], we evolve a population of 100 individuals SIT relations, initialized with Koza's 'Grow' method with program height set to 3. The upper limit on expression height in evolution is 8. Evolution lasts for 100 generations. All other parameters are as per the EpochX defaults.

We applied the above GP setup to all 35 unique letter strings occurring in the problems originally considered by Mitchell [27] (Sect. 10.2), i.e.:

abc	abd	ace	cab	cba	cde	cmg
cmz	edc	glz	ijk	kji	mrr	qbc
rst	rsu	xcg	xlg	xyz	aabc	aabd
abcd	abcm	abcn	ijkk	rijk	aababc	aabbcc
aabbcd	hhwwqq	iijjkk	lmfgop	mrrjjj	rssttt	xpqdef

and repeat each run ten times. On average, GP finds a correct SIT term (i.e., reproducing the target string perfectly, with $Lev(t, s) = 0$) in 93.4% of runs. For most strings, the success rate is 10/10, and the worst success rate is 4/10 (for `lmfgop`). The average information load amounts to 2.835, and the average number of nodes in a term is 7.190. GP managed to find SIT terms with minimal- or close-to minimal load for many problems, for instance:

- ijkk: $Group(Iter(i, Succ, 3), k)$
- aabbcc: $Iter(Group(a, a), Succ, 3)$
- xpqdef: $Seq(Group(x, Group(Iter(p, Succ, 2), Iter(d, Succ, 2))), f)$

Arguably, optimal SITs for these small problems could be found via exhaustive search. However, for more complex problems that we wish to handle prospectively, resorting to heuristic search is likely to be unavoidable.

10.4 Solving Proportional Analogies with GP

The CopyCat program [13] is a cognitive model of proportional analogy. Although very carefully engineered, the specifics of the interactions between its architectural elements (described in more detail in Sect. 10.7) are somewhat complex. While they have been described at length [13, 27], this is has nonetheless been done in a relatively informal fashion. It is therefore interesting to see if comparable results can be obtained by combining more readily-demarcated methods.

We therefore propose GPCAT, a GP-based method for tackling proportional analogies, with which we intend to achieve several goals:

Algorithm 1 Anti-unification algorithm for two terms

```
function AU(x,y)
    if x = y then
        return x
    else if x = f(x_1, ..., x_n) ∧ y = f(y_1, ..., y_n) then
        return f(AU(x_1, y_1), ..., AU(x_n, y_n))
    else
        return φ
    end if
end function
```

1. Compose well-known formalisms like SITs, Anti-Unification, and GP, rather than the mechanisms that are specific to CopyCat.
2. Verify GP's usefulness for solving proportional analogies.
3. Prospectively extend/substitute GPCAT's components with formalism for handling other domains more common to program synthesis.

There are three main components of GPCAT:

1. A domain-specific relational formalism (in this case SIT, Sect. 10.3).
2. An Anti-Unification algorithm.
3. A GP algorithm.

Anti-Unification (AU) is a procedure that extracts the common substructure of a set of terms T. The AU of T is itself a term, with some subterms replaced with *variables*. The defining property of such term u (*Anti-Unifier*) is that for each $t \in T$ there exists a *substitution* σ (i.e. a mapping from variables to terms) such that when applied to u, it makes it equal to t, i.e., $u\sigma = t$. In fact, u has the important property of being the most specific such term—informally, it preserves as much of the common structure as possible.

Algorithms for n-ary Anti-Unification were invented more-or-less simultaneously by Reynolds [34] and Plotkin [31]. For our purposes, unification of two terms (as per Algorithm 1) will suffice. The value ϕ denotes a so-called 'fresh variable', which maps to x under some substitution σ_x and to y under σ_y. The expressiveness of AU is dependent on how equality between terms is defined: in the case of *syntactic* AU that we consider here, function symbols are simply unique labels, with no intrinsic meaning.

Anti-Unification has been used in the solution of proportional analogy problems by Weller and Schmid [45]. Their algorithm is as follows [44]:

1. Use AU to compute the common structure of the terms A and C (Fig. 10.1), with associated substitutions σ_A, σ_C.
2. Determine D as $\sigma_C(\sigma_A^{-1}(B))$

For illustration, consider letter strings A=*abcg* and C=*ccbbaah*. Their natural representations in terms of SITs are respectively the following terms:

- $Seq(Iter(a, succ, 3), g)$
- $Seq(Iter(Group(c, c), pred, 3), h)$

The above algorithm returns the following AU of these terms: $Seq(Iter(\$1, \$2, 3), \$3)$ with substitutions $\sigma_A = \{\$1 \mapsto a, \$2 \mapsto succ, \$3 \mapsto g\}$ and $\sigma_C = \{\$1 \mapsto Group(c, c), \$2 \mapsto pred, \$3 \mapsto h\}$.

10.4.1 The GPCAT *Algorithm*

We now describe the application of GPCAT to the LSA domain. As we argue later, it can be also generalized to handle certain types of program synthesis problem. Given a proportional analogy problem, GPCAT generates a formal description of detected analogies/relationships, i.e., a set of expressions with variables, which can be then instantiated to generate the answers (i.e., the possible values for D). For some analogy of the form $A : B :: C : D$ (Fig. 10.1), GPCAT maintains a population of solutions, each of which is a *triple* of SIT terms, (t_A, t_B, t_C), intended to capture the respective structures for A, B, and C. The terms are subject to the same genetic search operators as in the single-term experiment presented in Sect. 10.3.1. The mutation operator randomly picks the term to be modified from t_A, t_B, and t_C; then, the selected term undergoes mutation as in Sect. 10.3.1, while the remaining two terms remain intact. Crossover operates analogously, i.e., the resulting offspring solutions diverge from the parents in only one of the terms.

The search goal is to synthesize a triple of SIT terms that not only reproduce the strings in LSA problem, but also together form a plausible analogical structure and ultimately yields the correct D. To this end, we attempt to capture the analogy between the horizontal and vertical mappings (h and v in Fig. 10.1) by performing Anti-Unification of their outcomes. As D is not given, the only explicitly known mappings are $h(t_A) = t_B$ and $v(t_A) = t_C$. These mappings share the same left-hand side t_A, so we perform Anti-Unification of their right-hand sides only, i.e., of t_B and t_C. This is also motivated by the fact that in most LSA problems, A plays the role of a mere 'anchor' for the symbols occurring in B and C; for instance in all but three LSA problems considered in [27], A is a sequence of three consecutive characters, typically `abc`.

We embed these computations into a fitness function which, for a given candidate solution (t_A, t_B, t_C), proceeds as follows:

1. Perform Anti-Unification of t_B and t_C to factor out their common substructure. This results in a term u with a number of variables $\$i$, $i = 1, \ldots, k$, and two substitutions σ_B and σ_C, such that $u\sigma_B = t_B$ and $u\sigma_C = t_C$. Technically, both σ_B and σ_C are sets of mappings from variables to subterms, e.g., $\sigma_B = \{\$1 \mapsto a, \$2 \mapsto Group(a, b)\}$. Symbols in right-hand sides of substitutions are represented as integer offsets w.r.t. the 'lowest' character occurring in the term (the importance of this will become clear in the example that follows).

2. Generate all 2^k combinations of mappings from σ_B and σ_C, resulting in 2^k 'artificial' substitutions $\sigma_j, j = 1, \ldots, 2^k$ (for low values of k in typical LSA problems, this can be done exhaustively).
3. Apply each σ_j independently to u, which results in a list of 2^k SIT terms. Express (i.e. 'flatten') the terms, obtaining so up to 2^k letter strings (distinct SIT terms, when expressed, may result in the same letter string). The resulting letter strings are the candidate answers, i.e., the proposed values of D, for the considered LSA problem.
4. Characterize the candidate solution (t_A, t_B, t_C) and the formal objects obtained in the above steps using following indicators:

   - $L = Lev(t_A, A) + Lev(t_B, B) + Lev(t_C, C)$, the total Levenshtein distance between expressed t_A, t_B, and t_C and respectively A, B and C; to be minimized (cf. *Lev* in Sect. 10.3.1).
   - $I = InfLoad(t_A) + InfLoad(t_B) + InfLoad(t_C)$, the total information load; to be minimized.
   - M, the total number of variables in u (equal also to the number of mappings in σ_B and σ_C); to be maximized, as the presence of multiple mappings may signal good structural correspondence of t_B and t_C.
   - N, the number of mappings to null value (i.e., $\$j \mapsto \epsilon$); to be minimized, as such mappings signal structural inconsistency between u and of the SIT terms it has been obtained from.

The indicators computed in step 4 form a multiobjective characterization of the evaluated candidate solution, and can be either aggregated into a single scalar fitness or handled by a multiobjective selection procedure. In this study, we follow the former option, and define minimized fitness as:

$$f((t_A, t_B, t_C)) = L + N + 0.01 * (I - M). \tag{10.2}$$

By taking into account several indicators, we mandate evolution to optimize all aspects of the analogy models simultaneously, i.e., conformance of SIT terms with the underlying LSA problem (L), low complexity of terms (I), and good Anti-Unification (M and N). Our fitness prioritizes L and N, i.e., puts solution correctness first.

Note that the proposed fitness function does not involve D, even if it is known. The correct D is expected to appear in the letter string list obtained in step 3.

We work through these steps for *abc : abd :: ijk : ?* (Fig. 10.1) and a candidate solution

$$(t_A, t_B, t_C) = (Iter(a, Succ, 3), Seq(Iter(a, Succ, 2), d), Iter(i, Succ, 3)).$$

Note that this solution reproduces all three terms perfectly, so its $L = 0$.

The anti-unifier of t_B and t_C (step 1 of GPCAT) calculated using first-order, rigid, unranked AU algorithm [2], is given by:

$$Seq(Iter(\$1, succ, \$2), \$3),$$

with $\sigma_B = \{\$1 \mapsto a, \$2 \mapsto 2, \$3 \mapsto d\}$, and $\sigma_C = \{\$1 \mapsto i, \$2 \mapsto 3, \$3 \mapsto \epsilon\}$. Now, as signalled in Step 1 of GPCAT, the symbols in right-hand sides of substitutions are represented as offsets w.r.t. the lowest characters (here a and i, respectively), so the substitutions take the following form (note the underlined differences): $\sigma_B = \{\$1 \mapsto \underline{0}, \$2 \mapsto 2, \$3 \mapsto \underline{3}\}$, and $\sigma_C = \{\$1 \mapsto \underline{0}, \$2 \mapsto 3, \$3 \mapsto \epsilon\}$. With $k = 3$ variables, there are $2^3 = 8$ artificial substitutions σ_j that can be built by combining the individual mappings from σ_B and σ_C (step 2 of GPCAT). Among them, there is $\sigma_3 = \{\$1 \mapsto \underline{0}, \$2 \mapsto 2, \$3 \mapsto \underline{3}\}$, which for initial character i produces ijl, the most natural answer to this LSA problem.

10.5 The Experiment

We applied GPCAT to 32 out of 35 LSA problems originally considered by Mitchell [27], i.e., those problems with A being a sequence of three consecutive letters. Instruction set (SIT operators) and evolutionary parameters were set as in Sect. 10.3.1, except for higher initial tree height (5, to promote diversity in initial population) and lower than usual selection pressure (tournament of size 2, in order to promote exploration and lower the risk of premature convergence). This time we relied on implementation based on the FUEL evolutionary computation library written in Scala.[1]

The best-of-run solutions resulting from particular runs were subject to evaluation, and the lists of answers to the problem (i.e., element 'D' in $A : B :: C : D$) was collected with 30 runs for each LSA problem. Table 10.1 presents the top five most frequently occurring answer strings per 30 runs of GPCAT for selected problems from the considered suite. Each string is accompanied with the percentage of times it has occurred. By contrast, CopyCat responses [27], shown in Table 10.2 (also as per cents of runs), do not sum up to 100%, as a CopyCat run produces a single answer. GPCAT's outcomes tend to only partially coincide with those of CopyCat: for instance for the first problem, *ijl* is the most common answer in both methods, while for the second problem their outcomes do not overlap at all (one

Table 10.1 The top five most frequently occurring answer strings per 30 runs of GPCAT

Problem	Most frequent answers				
abc:abd::ijk	ijl:100	ik:7	bcd:7	abbd:7	ac:7
abc:abd::xyz	xya:100	bcd:7	abbd:7	xz:0.07	ac:7
abc:abd::kji	ijl:70	cba:57	kln:17	bce:10	jl:7
abc:qbc::iijjkk	aabbcc:53	ijl:43	ab:23	ij:23	ik:10
abc:abd::mrrjjj	jkm:67	iiaaa:33	rrjjj:33	jrrjjj:17	diiaaa:17

[1] https://github.com/kkrawiec/fuel.

Table 10.2 The top five most frequently occurring answer strings of GPCAT

Problem	Most frequent answers				
abc:abd::ijk	ijl:96.9	ijd:2.7	ijk:0.2	hjk:0.1	ijj:0.1
abc:abd::xyz	xyd:81.1	wyz:11.4	yyz:6	dyz:0.7	xyz:0.4
abc:abd::kji	kjh:56.1	kjj:23.8	lji:18.6	kjd:1.1	kki:0.3
abc:abd::iijjkk	iijjll: 81.0	iijjkl: 16.5	iijjdd: 0.9	iikkll: 0.9	iijkll: 0.3
abc:abd::mrrjjj	mrrkkk:70.5	mrrjjk:19.7	mrrjkk:4.8	mrrjjjj:4.2	mrrjjd:0.6

of the reasons being that GPCAT's process of variable alignment has built-in the concept of modulo, whereas by design, CopyCat's domain knowledge excludes a successor to 'z'). One possible research direction is thus tweaking and extending GPCAT in order to match the distribution of human answers (of which CopyCat, despite being concerned with plausible solutions rather than slavish reproduction of human bias, is arguably the best known computational model).

However, exact mimicking of human behaviour, though interesting from the viewpoint of cognitive science, might be of lesser importance for program synthesis. What might be more essential in the latter context is the sole concept of proportional analogy, and generative mechanism for their creation based on structural Anti-Unification. We discuss this perspective in the following section.

10.6 Analogies in Program Synthesis

Let us now illustrate why we find analogical reasoning a useful concept for test-based program synthesis. Consider the domain of list manipulation and the task of synthesizing the `append` function. Let the desired behaviour of that function be specified by the following set of tests:

```
append([1,2], [])    = [1,2]
append([1,2], [3])   = [1,2,3]
append([1,2,3], [])  = [1,2,3]
append([a,b], [c])   = [a,b,c]
```

By selecting pairs of tests from this list, we may form the following proportional analogies:

$$\begin{array}{ccc} ([1,2],[3]) & \longrightarrow & [1,2,3] \\ \downarrow & & \downarrow \\ ([a,b],[c]) & \longrightarrow & [a,b,c] \end{array} \qquad \begin{array}{ccc} ([1,2],[]) & \longrightarrow & [1,2] \\ \downarrow & & \downarrow \\ ([1,2,3],[]) & \longrightarrow & [1,2,3] \end{array} \qquad \begin{array}{ccc} ([1,2],[3]) & \longrightarrow & [1,2,3] \\ \downarrow & & \downarrow \\ ([1,2,3],[]) & \longrightarrow & [1,2,3] \end{array}$$

These analogies capture three unrelated characteristics of the synthesis task. The first one is type-related and says that `append` takes no notice of the nature of the list elements: in a sense, it behaves 'modulo' type, whether list elements are characters or numbers. The second analogy concerns more the operational characteristics of `append`, and signals that if the second argument of `append` is an empty list, then the expected result is the first argument. The third analogy might be seen as expressing an invariant; i.e. that moving the head of the second list to the end of the first list does not change the outcome.

On the face of it, these analogies express quite trivial facts. Nevertheless, our case in point is that just by juxtaposing existing tests (i.e., without reaching to any source of extra knowledge), we obtain concepts that capture *various qualities of desired behaviour*. We claim that (1) identification of such qualities and (2) their separation can make program synthesis more efficient. Conventional GP has all these test cases at its disposal, yet is completely oblivious to this opportunity.

We believe that this can provide a basis for the induction of high-level, 'global' descriptions of a set of fitness cases from repeated encounters with local ones by the search process [18]. This then begs the wide-ranging research question of how to exploit such induced invariants for use as *search drivers* [19, 20, 22], i.e., additional quasi-objectives that guide the search process. Depending on the domain, it may be possible to express them as predicates in the same function set as is used to solve the problem. Alternatively, it may be desirable to add induced invariants to a competitive co-evolutionary population of constraints. In either case, our approach yields relational linkages in a functional, hierarchical manner, as opposed to the traditional models of relational linkage occasionally used in stochastic program induction, which are primarily probabalistic [46].

In a broader perspective, of particular interest here is the prospect of using the generative aspects of GP to help address a persistent problem in formal methods. As observed by Luqi and Goguen [24], *"formal methods tend to be brittle or discontinuous—a small change in the domain can require a great deal of new work"*. Since formal approaches can be sensitive to the particular manner in which their input is presented, the ability to generate alternative representations for inputs may bring benefits not available to either approach in isolation. Conversely, it was observed by Kocsis and Swan that the formal structure of inductively-defined datatypes can usefully be exploited for GP purposes, e.g. to eliminate otherwise stochastic operations [17]. We might also hope to make use of this kind of structure for our current purposes. For example, it is well-known that lists (and indeed algebraic datatypes in general) can be expressed in a relational manner, in this case via the type constructors Nil and Cons. Hence [1,2,3] can be expressed as:

$$Cons(1, Cons(2, Cons(3, Nil))).$$

Using SIT-style relations, this can be represented as `Iter(Nil,succ,3)`,[2] and the second analogy can be represented by the Anti-Unifier:

$$App(Iter(Nil, succ, \$1), Nil), Iter(Nil, succ, \$1)),$$

[2]Strictly, `Iter` here is slightly more complex than that previously mentioned, in that it expresses an inductive construction known as a *catamorphism* [26].

with substitutions $\sigma_1 = \{\$1 \mapsto 2\}$, $\sigma_2 = \{\$1 \mapsto 3\}$, in congruence with the fact that appending Nil preserves structure.

Finally, we note that while the 'mixing' properties of binary recombination have been widely examined in the EC community, even though Yu notes that structure abstraction can contribute to success [47], the notion of an 'abstracting' binary operator has not, to our knowledge, been further explored. It would therefore be interesting to consider generalization of two programs as an addition to the traditional palette of binary recombination operators.

10.7 Related Work

Foundational computational work in proportional analogy was done by van der Helm and Leeuwenberg [42], describing the problem in terms of path search in directed acyclic graphs and giving an algorithm which is $\mathcal{O}(n^4)$ in the size of the input. This was subsequently extended by Dastani et al. [4] to incorporate the algebraic approach to SIT adopted in this article. Dastani also applied GP (with an uncharacteristically high mutation rate of 0.4) to the induction of SIT structures for linear line patterns [5], i.e. polylines which can be encoded as letterstrings.

CopyCat [13] is perhaps the most well-known architecture for solving proportional analogies. It has a tripartite structure, consisting of a blackboard ('the workspace'), a priority queue of programs for updating blackboard state ('the coderack') and a semantic network with dynamically re-weighted link strengths ('the slipnet'). CopyCat is entirely concerned with (predominantly local) mechanisms that have cognitive plausibility.

Closest to the current work is the algorithm of Weller and Schmid [45] for solving proportional analogies, which performs anti-unification via E-generalization. The representation for E-generalization is a regular tree grammar, which means that the result is a (potentially infinite) equivalence class of terms for D. The claimed advantages for their approach are twofold:

1. There is no need to explicitly induce SIT representations for A, B, C, since all are represented simultaneously via the regular tree grammar.
2. All consistent values for D are likewise represented simultaneously.

However, this approach suffers from the severe disadvantage that no mechanism is provided for enumerating the resulting regular tree grammar in preference order (e.g. by information load). Since it is not possible to distinguish certain specific representations for D as being more compelling, it also does not appear to be of practical use. In contrast, our approach induces SIT representations with low information load via GP driven by multi-aspect fitness function, then uses syntactic AU to determine D.

Early use of analogical mechanisms for program synthesis predominantly operated on specifications rather than concrete programs [6, 7, 25, 41]. More recently, Schmid learned programs from fitness cases via planing [35] and Raza et al. [33]

used Anti-Unification to address scalability issues in synthesising DSL programs for XML transformation.

IGOR II [12] is currently considered the exemplar of program synthesis by Inductive Functional Programming (IFP) [28]. It creates a recursive program to generalize a set of fitness cases via a pattern-based rewriting system, having first obtained the least general generalization of the set of fitness cases by examples by AU. Katayama [16] categorized approaches to IFP into analytical approaches based on analysis of fitness cases and generate-and-test approaches that create many candidate programs. The IGOR II algorithm is further extended by Katayama to hybridize these two approaches.

10.8 Conclusions

In this chapter, we discussed two-way liaisons between GP-based program synthesis and analogical reasoning. We showed that, on one hand, GP can be employed to solve proportional analogy problems with aid of structural representations (SIT terms) and a formal Anti-Unification mechanism. On the other hand—and more importantly—we pointed out to potential ways of improving the efficiency of a GP search process via detection and structural characterization of analogies between fitness cases.

In this study, we have only scratched the surface regarding the exploitation of analogical reasoning for GP-based program synthesis. For example, we have limited our attention to analogies built on *pairs* of tests. Arguably, other interesting and potentially useful structures could be obtained by working with multiple tests at a time. We hypothesize that one way of attaining this goal could be via hierarchically aggregating analogies, i.e., forming analogies of the form `Case1 : Case2 :: Case3 : Case4`. Another possibility is to exploit the knowledge captured by analogies for parent (mate) selection: arguably, two programs that happen to 'solve' analogies based on different pairs of tests feature complementary characteristics that may be worth combining. These observations point to next steps in the research agenda of analogy-based programming.

Acknowledgements Thanks are due to Dave Bender and the CRCC in Bloomington for providing us with the original list of letter-string analogy examples. K. Krawiec acknowledges support from grant 2014/15/B/ST6/05205 funded by the National Science Centre, Poland. Both authors thank the reviewers for valuable and insightful suggestions and comments.

References

1. Aamodt, A., Plaza, E.: Case-based reasoning: foundational issues, methodological variations, and system approaches. AI Commun. **7**(1), 39–59 (1994). http://dl.acm.org/citation.cfm?id=196108.196115

2. Baumgartner, A., Kutsia, T.: A Library of Anti-Unification Algorithms. RISC Report Series 14-07, Research Institute for Symbolic Computation (RISC), Johannes Kepler University Linz, Schloss Hagenberg, Hagenberg (2014). http://www.risc.jku.at/publications/download/risc_5003/au_library.pdf
3. Cooke, H., Tredennick, H.: Aristotle: the Organon, vol. 1. Harvard University Press, Harvard (1938). https://books.google.co.uk/books?id=TgeISwAACAAJ
4. Dastani, M., Indurkhya, B., Scha, R.: Analogical projection in pattern perception. J. Exp. Theor. Artif. Intell. **15**(4), 489–511 (2003). https://doi.org/10.1080/09528130310001626283
5. Dastani, M., Marchiori, E., Voorn, R.: Finding perceived pattern structures using genetic programming. In: Spector, L., et al. (eds.) Proceedings of the Genetic and Evolutionary Computation Conference (GECCO-2001), pp. 3–10. Morgan Kaufmann, San Francisco, CA (2001). http://www.cs.bham.ac.uk/~wbl/biblio/gecco2001/d01.pdf
6. Dershowitz, N.: The evolution of programs: program abstraction and instantiation. In: Proceedings of the 5th International Conference on Software Engineering, ICSE '81, pp. 79–88. IEEE Press, Piscataway, NJ (1981). http://dl.acm.org/citation.cfm?id=800078.802519
7. Dershowitz, N., Manna, Z.: On automating structured programming. In: Huet G., Kahn, G. (eds.) IRIA Symposium on Proving and Improving Programs, pp. 167–193. Arc-et-Senans (1975)
8. Ehrenfels, C.V.: Über Gestaltqualitäten. Vierteljahresschr. für Philosophie **14**, 249–292 (1890)
9. Evans, T.G.: A heuristic program to solve geometric-analogy problems. In: Proceedings of the April 21–23, 1964, Spring Joint Computer conference, AFIPS '64 (Spring), pp. 327–338. ACM, New York, NY (1964). http://doi.acm.org/10.1145/1464122.1464156
10. Falkenhainer, B., Forbus, K.D., Gentner, D.: The structure-mapping engine: algorithm and examples. Artif. Intell. **41**(1), 1–63 (1989). http://dx.doi.org/10.1016/0004-3702(89)90077-5
11. French, R.M.: The subtlety of sameness: a theory and computer model of analogy-making. The MIT Press, Cambridge (1995)
12. Hofmann, M.: Igor II - an analytical inductive functional programming system. In: In Proceedings of the 2010 ACM SIGPLAN Workshop on Partial Evaluation and Program Manipulation, pp. 29–32 (2010)
13. Hofstadter, D.R.: Fluid Concepts and Creative Analogies: Computer Models of the Fundamental Mechanisms of Thought. Basic Books, Inc., New York, NY (1996)
14. Holyoak, K.J., Thagard, P.: Analogical mapping by constraint satisfaction. Cogn. Sci. **13**(3), 295–355 (1989). http://dx.doi.org/10.1207/s15516709cog1303_1
15. Hummel, J.E., Holyoak, K.J.: Distributed representations of structure: a theory of analogical access and mapping. Psycholog. Rev. **1997**, 427–466 (1997)
16. Katayama, S.: An analytical inductive functional programming system that avoids unintended programs. In: Proceedings of the ACM SIGPLAN 2012 Workshop on Partial Evaluation and Program Manipulation, PEPM '12, pp. 43–52. ACM, New York, NY (2012). http://doi.acm.org/10.1145/2103746.2103758
17. Kocsis, Z.A., Swan, J.: Asymptotic Genetic Improvement programming via type functors and catamorphisms. In: Johnson, C., Krawiec, K., Moraglio, A., O'Neill, M. (eds.) Semantic Methods in Genetic Programming. Ljubljana, Slovenia (2014). http://www.cs.put.poznan.pl/kkrawiec/smgp2014/uploads/Site/Kocsis.pdf. Workshop at Parallel Problem Solving from Nature 2014 Conference
18. Kovitz, B., Swan, J.: Structural stigmergy: a speculative pattern language for metaheuristics. In: Proceedings of the Companion Publication of the 2014 Annual Conference on Genetic and Evolutionary Computation, GECCO Comp '14, pp. 1407–1410. ACM, New York, NY (2014). http://doi.acm.org/10.1145/2598394.2609845
19. Krawiec, K.: Behavioral Program Synthesis with Genetic Programming, 1st edn. Springer Publishing Company, Incorporated, Berlin (2015)
20. Krawiec, K., O'Reilly, U.M.: Behavioral programming: a broader and more detailed take on semantic GP. In: Proceedings of the 2014 Annual Conference on Genetic and Evolutionary Computation, GECCO '14, pp. 935–942. ACM, New York, NY (2014). http://doi.acm.org/10.1145/2576768.2598288

21. Krawiec, K., Swan, J.: Guiding evolutionary learning by searching for regularities in behavioral trajectories: a case for representation agnosticism. In: AAAI Fall Symposium: How Should Intelligence be Abstracted in AI Research (2013)
22. Krawiec, K., Swan, J., O'Reilly, U.M.: Behavioral program synthesis: Insights and prospects. In: Riolo, R., Worzel, W.P., Groscurth, K. (eds.) Genetic Programming Theory and Practice XIII, Genetic and Evolutionary Computation. Springer, Ann Arbor (2015). http://www.cs.put.poznan.pl/kkrawiec/wiki/uploads/Research/2015GPTP.pdf
23. Leeuwenberg, E., van der Helm, P.: Structural Information Theory: The Simplicity of Visual Form. Cambridge University Press, Cambridge (2015)
24. Luqi, Goguen, J.A.: Formal methods: promises and problems. IEEE Softw. **14**(1), 73–85 (1997). http://dx.doi.org/10.1109/52.566430
25. Manna, Z., Waldinger, R.: Knowledge and reasoning in program synthesis. In: Programming Methodology, 4th Informatik Symposium, pp. 236–277. Springer, London (1975). http://dl.acm.org/citation.cfm?id=647950.742874
26. Meijer, E., Fokkinga, M., Paterson, R.: Functional programming with bananas, lenses, envelopes and barbed wire. In: Proceedings of the 5th ACM Conference on Functional Programming Languages and Computer Architecture, pp. 124–144. Springer New York, Inc., New York (1991). http://dl.acm.org/citation.cfm?id=127960.128035
27. Mitchell, M.: Analogy-Making as Perception: A Computer Model. MIT Press, Cambridge (1993). http://portal.acm.org/citation.cfm?id=152203
28. Muggleton, S.: Inductive Logic Programming: Derivations, successes and shortcomings. SIGART Bull. **5**(1), 5–11 (1994). http://doi.acm.org/10.1145/181668.181671
29. Otero, F., Castle, T., Johnson, C.: EpochX: Genetic programming in java with statistics and event monitoring. In: Proceedings of the 14th Annual Conference Companion on Genetic and Evolutionary Computation, GECCO '12, pp. 93–100. ACM, New York, NY (2012). https://doi.org/10.1145/2330784.2330800
30. Phillips, S., Wilson, W.H.: Categorial compositionality: a category theory explanation for the systematicity of human cognition. PLoS Comput. Biol. **6**(7) (2010)
31. Plotkin, G.D.: A note on inductive generalization. Mach. Intell. **5**, 153–163 (1970)
32. Prade, H., Richard, G.: Computational Approaches to Analogical Reasoning: Current Trends. Springer Publishing Company, Incorporated, Berlin (2014)
33. Raza, M., Gulwani, S., Milic-Frayling, N.: Programming by example using least general generalizations. In: Proceedings of the Twenty-Eighth AAAI Conference on Artificial Intelligence, July 27–31, 2014, Québec City, QC, pp. 283–290 (2014). http://www.aaai.org/ocs/index.php/AAAI/AAAI14/paper/view/8520
34. Reynolds, J.C.: Transformational systems and the algebraic structure of atomic formulas. In: Meltzer, B., Michie, D. (eds.) Machine Intelligence, vol. 5, pp. 135–151. Edinburgh University Press, Edinburgh (1969)
35. Schmid, U.: Inductive Synthesis of Functional Programs, Universal Planning, Folding of Finite Programs, and Schema Abstraction by Analogical Reasoning, Lecture Notes in Computer Science, vol. 2654. Springer, Berlin (2003). http://dx.doi.org/10.1007/b12055
36. Schmid, U., Burghardt, J.: An algebraic framework for solving proportional and predictive analogies. In: Schmalhofer, F., et.al. (eds.) Proceedings of the European Conference on Cognitive Science, pp. 295–300, Erlbaum (2003)
37. Schmidt, M., Krumnack, U., Gust, H., Kühnberger, K.: Heuristic-driven theory projection: an overview. In: Prade, H., Richard, G. (eds.) Computational Approaches to Analogical Reasoning: Current Trends, vol. 548, pp. 163–194. Springer, Berlin (2014). https://doi.org/10.1007/978-3-642-54516-0_7
38. Silva, S., Dignum, S., Vanneschi, L.: Operator equalisation for bloat free genetic programming and a survey of bloat control methods. Genet. Program Evolvable Mach. **13**(2), 197–238 (2012). http://dx.doi.org/10.1007/s10710-011-9150-5
39. Stewart, I., Cohen, J.: Figments of Reality: The Evolution of the Curious Mind. Cambridge University Press, Cambridge (1999)

40. Swan, J., Drake, J., Krawiec, K.: Semantically-meaningful numeric constants for genetic programming. In: Johnson, C., Krawiec, K., Moraglio, A., O'Neill, M. (eds.) Semantic Methods in Genetic Programming. Ljubljana, Slovenia (2014). http://www.cs.put.poznan.pl/kkrawiec/smgp2014/uploads/Site/Swan.pdf. Workshop at Parallel Problem Solving from Nature 2014 Conference
41. Ulrich, J.W., Moll, R.: Program synthesis by analogy. SIGART Bull. **64**, 22–28 (1977). http://doi.acm.org/10.1145/872736.806928
42. van der Helm, P., Leeuwenberg, E.: Avoiding explosive search in automatic selection of simplest pattern codes. Pattern Recogn. **19**(2), 181–191 (1986). http://dx.doi.org/10.1016/0031-3203(86)90022-1
43. Van Der Helm, P.A., Leeuwenberg, E.L.J.: Accessibility: A criterion for regularity and hierarchy in visual pattern codes. J. Math. Psychol. **35**(2), 151–213 (1991). http://dx.doi.org/10.1016/0022-2496(91)90025-O
44. Weller, S., Schmid, U.: Analogy by abstraction. In: Proceedings of the Seventh International Conference on Cognitive Modeling (ICCM). Trieste (2006)
45. Weller, S., Schmid, U.: Solving proportional analogies by *E*-generalization. In: KI 2006: Advances in Artificial Intelligence, 29th Annual German Conference on AI, KI 2006, Bremen, Germany, June 14–17, 2006, Proceedings, pp. 64–75 (2006). https://doi.org/10.1007/978-3-540-69912-5_6
46. Yanai, K., Iba, H.: Estimation of distribution programming: EDA-based approach to program generation. In: Towards a New Evolutionary Computation: Advances in the Estimation of Distribution Algorithms, pp. 103–122. Springer, Berlin (2006). https://doi.org/10.1007/3-540-32494-1_5
47. Yu, T.: Structure abstraction and genetic programming. In: Angeline, P.J., Michalewicz, Z., Schoenauer, M., Yao, X., Zalzala, A. (eds.) Proceedings of the Congress on Evolutionary Computation, vol. 1, pp. 652–659. IEEE Press, Mayflower Hotel, Washington, DC (1999). https://doi.org/10.1109/CEC.1999.781995. http://www.cs.mun.ca/~tinayu/index_files/addr/public_html/cec99.pdf

Chapter 11
An Evolutionary Algorithm for Big Data Multi-Class Classification Problems

Michael F. Korns

Abstract As symbolic regression (SR) has advanced into the early stages of commercial exploitation, the poor accuracy of SR still plagues even advanced commercial packages, and has become an issue for industrial users. Users expect a correct formula to be returned, especially in cases with zero noise and only one basis function with minimal complexity. At a minimum, users expect the response surface of the SR tool to be easily understood, so that the user can know *a priori* on what classes of problems to expect excellent, average, or poor accuracy. Poor or *unknown* accuracy is a hindrance to greater academic and industrial acceptance of SR tools. In several previous papers, we presented a complex algorithm for modern SR, which is extremely accurate for a large class of SR problems on noiseless data. Further research has shown that these extremely accurate SR algorithms also improve accuracy in noisy circumstances—albeit not *extreme* accuracy. Armed with these SR successes, we naively thought that achieving extreme accuracy applying GP to symbolic multi-class classification would be an easy goal. However, it seems algorithms having extreme accuracy in SR do not translate directly into symbolic multi-class classification. Furthermore, others have encountered serious issues applying GP to symbolic multi-class classification (Castelli et al. Applications of Evolutionary Computing, EvoApplications 2013: EvoCOMNET, EvoCOMPLEX, EvoENERGY, EvoFIN, EvoGAMES, EvoIASP, EvoINDUSTRY, EvoNUM, EvoPAR, EvoRISK, EvoROBOT, EvoSTOC, vol 7835, pp 334–343. Springer, Vienna, 2013). This is the first paper in a planned series developing the necessary algorithms for extreme accuracy in GP applied to symbolic multi-class classification. We develop an evolutionary algorithm for optimizing a single symbolic multi-class classification candidate. It is designed for big-data situations where the computational effort grows linearly as the number of features and training points increase. The algorithm's behavior is demonstrated on theoretical problems, UCI benchmarks, and industry test cases.

M. F. Korns (✉)
Analytic Research Foundation, Henderson, NV, USA

R. Riolo et al. (eds.), *Genetic Programming Theory and Practice XIV*, Genetic and Evolutionary Computation,
https://doi.org/10.1007/978-3-319-97088-2_11

Keywords Genetic programming · Symbolic classification · Particle swarm · Abstract expression grammar · Grammar template genetic programming · Genetic algorithms

11.1 Introduction

The discipline of Genetic Programming (GP) [12–14] has matured significantly in the last two decades. There are numerous practical successes reported in many application domains [2, 19]. A great deal of work has been done to strengthen the theoretical foundations of GP [15]. There is at least one commercial package Symbolic Regression (SR) package which has been on the market for several years (http://www.rmltech.com/). There is now at least one well documented commercial symbolic regression package available for Mathematica (http://www.evolved-analytics.com). There is at least one very well done open source symbolic regression package available for free download (http://ccsl.mae.cornell.edu/eureqa). In addition to our own ARC system [5], currently used internally for massive (million row) financial data nonlinear regressions, there are a number of other mature symbolic regression packages currently used in industry including [20] and [11]. Plus there is another commercially deployed regression package which handles up to 50 to 10,000 input features using specialized linear learning [16].

Yet, despite its increasing sophistication, genetic programming has encountered serious issues addressing multi-class classification applications [1].

An algorithm has been developed, specifically for genetic programming applications in multi-class classification, called M2GP, which has achieved reasonable accuracy in several multi-class classification tests [3]. One highly attractive attribute of the M2GP algorithm is its solid theoretical machine learning foundations. One unfortunate issue with the M2GP algorithm, is the requirement to compute several matrix multiplications and one matrix inversion per class, and the necessity that the crucial class covariance matrix be non-singular. Furthermore, the M2GP algorithm requires several vector products and one matrix multiply per training point for scoring. As the number of classes and the number of training examples grows larger, in a multi-class classification problem, the computational requirements for the M2GP algorithm increases geometrically. In financial applications there is no a priori guarantee that all of the class covariance matrices will be non-singular. Plus, at least in many financial applications, we have found that the number of classes and training examples can be quite large.

We are interesting in exploring a purely evolutionary alternative to the M2GP algorithm which will not require matrix inversions, and which will operate correctly in conditions wherein the M2GP class covariance matrix is singular.

Before continuing with the discussion of our alternative multi-class classification algorithms for big data problems, we proceed with a basic introduction and formalization of Genetic Programming Classification (GPC), liberally adapting terminology from M2GP [1] and from Generalized Linear Models (GLMs) [17].

The formalization of genetic programming classification is the class of Genetic Programming Classifier Models (GPCMs). A GPCM is a collection of K discriminant functions $D^k, k \in \{1, 2, \ldots, K\}$ a dependent unordered categorical variable y with integer values from 1 through K, and an independent data point with M features $x = (x_1, \ldots, x_M)$, such that

$$ey(x) = nomial(c_0^1 + c_1^1 \times D^1(x) + \cdots + c_0^K + c_1^K \times D^K(x)), \tag{11.1}$$

$$a_k = max(a_1, \ldots, a_K), \text{ implies } k = nomial(a_1, \ldots, a_K)), \tag{11.2}$$

$$y = ey \text{ implies } match(y, ey) = 0 \text{ and } y \neq ey, \text{ implies } match(y, ey) = 1. \tag{11.3}$$

Given a collection of GPCMs, a collection of independent data points, X, and a collection of dependent categorical variables, Y, the fittest GPCM is the GPCM which minimizes $match(Y, EY)$.

The discriminant functions are a broad generalization and can represent any possible linear or nonlinear formula, as in the following examples:

$$D^1 = x_3 \tag{11.4}$$

$$D^2 = x_1 + x_4 \tag{11.5}$$

$$D^3 = sqrt(x_2) \div tan(x_5 \div 4.56) \tag{11.6}$$

$$D^4 = tanh(cos(x_2 \times 0.2) \times cube(x_5 + abs(x_1))) \tag{11.7}$$

Viewing the problem in this fashion, we gain an important insight. Genetic programming classification does not add anything to the standard techniques of classification. The value added by GPC lies in its abilities as a search technique: how quickly and how accurately can GPC find an optimal set of discriminant functions $\{D\}$. The immense size of the search space provides ample need for improved search techniques. In basic Koza-style tree-based Genetic Programming [12] the genome and the individual are the same Lisp s-expression which is usually illustrated as a tree. Of course the tree-view of an s-expression is a visual aid, since a Lisp s-expression is normally a list which is a special Lisp data structure. Without altering or restricting basic tree-based GP in any way, we can view the individual discriminant functions not as trees but instead as s-expressions such as this depth 2 binary tree s-expression: `(/ (+ x_2 3.45) (* x_0 x_2))`, or this depth 2 irregular tree s-expression: `(/ (+ x_4 3.45) 2.0)`.

In basic GP the non-terminal nodes are all operators (implemented as Lisp function calls), and the terminal nodes are always either real number constants or input features. The maximum depth of a GP individual is limited by the available computational resources; but, it is standard practice to limit the maximum depth of a GP individual to some manageable limit at the start of a genetic programming run.

Given any selected maximum depth d, it is an easy process to construct a maximal binary tree s-expression U_d, which can be produced by the GP system without violating the selected maximum depth limit. As long as we are reminded that each f represents a function node while each t represents a terminal node (either a feature v or a real number constant c), the construction algorithm is simple and recursive as follows:

(U_0) `t`
(U_1) `(f t t)`
(U_2) `(f (f t t) (f t t))`
(U_3) `(f (f (f t t) (f t t)) (f (f t t) (f t t)))`
(U_d) `(f` U_{d-1} U_{d-1}`)`

The basic GP symbolic regression system [12], which we will adapt for symbolic classification, contains a set of functions F, and a set of terminals T. If we let $t \in T$, and $f \in (F \cup \xi)$, where $\xi(a, b) = \xi(a) = a$, then any basis function produced by the basic GP system will be represented by at least one element of U_d. Adding the ξ function allows U_d to express all possible basis functions generated by the basic GP system *to a depth of d*.

To emphasize this important point: The ξ function performs the job of a pass-through function. The ξ function allows a *fixed-maximal-depth* expression in U_d to express trees of varying depth, such as might be produced from a GP system. For instance, the varying depth GP expression $x_2 + (x_3 - x_5) = \xi(x_2, 0.0) + (x_3 - x_5) = \xi(x_2, 0.0) - (x_3 x_5)$ which is a *fixed-maximal-depth* expression in U_2.

In addition to the special pass through function ξ, in our system we also make additional slight alterations to improve coverage, reduce unwanted errors, and restrict results from wandering into the complex number range. All unary functions, such as cos, are extended to ignore any extra arguments so that, for all unary functions, $cos(a, b) = cos(a)$. The $sqrt$ and ln functions are extended for negative arguments so that $sqrt(a) = sqrt(|a|)$ and $ln(a) = ln(|a|)$.

Given this formalism of the search space, it is easy to compute the size of the search space, and it is easy to see that the search space is huge even for rather simple discriminant functions. For our use in this chapter the function set will be the following functions: $F = \{+, -, *, /, \leq, \geq, max, min, inv, \xi\}$, where $inv(x) = (1.0/x)$, $(x \leq y)$ is 1.0 if true or 0.0 if false, $(x \geq y)$ is 1.0 if true or 0.0 if false, and $\xi(a, b) = \xi(a) = a$. The terminal set are the features x_1 through x_M, and the real constant c, which we shall consider to be 2^{64} in size.

The fitness measure used in this paper is the Classification Error Percent (CEP) computed as the average error percent on a per class basis. The formula for the computation of the CEP fitness measure is shown in fitness equation (11.8). The term $Error_k$ refers to the total number of unmatched cases for the class k—that is, $match(ey, y) = 1$ for class k. The term $Count_k$ refers to the total number of training points for class k.

$$\text{CEP} = average(\text{Error}_k \div \text{Count}_k) \tag{11.8}$$

In this paper we introduce a Multilayer Discriminant Classifier (MDC) algorithm for big data genetic programming multi-class classification. The MDC algorithm is evolutionary in approach. Each layer of the algorithm requires only one pass over the data, *including scoring*. So MDC's computational complexity grows linearly with larger training data points and classes. Furthermore, MDC's multilayer approach distributes the evolutionary attention across the entire genetic programming run so that more promising candidates receive more evolutionary attention and less promising candidates receive less evolutionary attention.

Prior to writing this chapter, a great deal of *tinker-engineering* was performed on the Lisp code supporting the MDC algorithm. For instance, all generated candidate code was checked to make sure that the real numbers were loaded into Intel machine registers without exception. All vector pointers were checked to make sure they were loaded into Intel address registers at the start of each loop rather than re-loaded with each feature reference. As a result of these engineering efforts, the MDC algorithm is quite practical to run on a personal computer. Of course a cloud configuration can always be used to achieve enhanced performance in much shorter elapsed times.

11.2 The MDC Algorithm

The Multilayer Discriminant Classification (MDC) algorithm is composed of three main layers of evolutionary activity. These three layers focus on the optimization of the GPCM coefficients for the optimal GPCM candidate, and are a direct alternative to the more mathematically correct M2GP algorithm. None of the MDC algorithm layers require matrix inversion. Therefore the MDC algorithm can operate in training conditions which do not meet the basic requirements of the M2GP algorithm.

The MDC algorithm attempts to optimize the following equation for the K selected discriminant functions $D^k(x)$:

$$ey(x) = nomial(c_0^1 + (c_1^1 \times D^1(x)), \ldots, c_0^K + (c_1^K \times D^K(x))). \tag{11.9}$$

The MDC algorithm is evolutionary in approach. Each layer of the algorithm requires only one pass over the data, *including scoring*. The MDC algorithm's multilayer approach distributes the evolutionary attention across the entire genetic programming run so that more promising candidates receive more evolutionary attention and less promising candidates receive less evolutionary attention.

11.2.1 Partial Bipolar Regression

The first layer of the MDC algorithm attempts to find a quick initial approximation of the optimal coefficients in the selected GPCM.

$$ey(x) = nomial(c_0^1 + (c_1^1 \times D^1(x)), \ldots, c_0^K + (c_1^K \times D^K(x)))$$

At initialization time, randomly selected coefficients for the current candidate GPCM (11.1) will generally score a Classification Error Percent (CEP) between 80% and 99% especially when as K grows larger *even when the GPCM is an exact match for the dependent variable Y*. We need a quick initialization approach which scores a CEP much closer to 0% even for large K. Partial Bipolar Regression (PBR) is just such a quick GPCM coefficient initialization methodology*especially when the GPCM is an exact match for the dependent variable Y*.

For each new GPCM, the MDC algorithm runs K Partial Bipolar Regression (PBR) single passes through the data followed by a single scoring pass thought the data. To understand how this works, notice that formula (11.1) is composed of K simple discriminant formulas as follows:

$$c_0^k + (c_1^k \times D^k(x)). \tag{11.10}$$

The PBR layer runs a simple single pass regression on each of the K discriminant formulas against the dependent variable y; however, as each y is loaded its value is temporarily altered on the fly according to the following rule. If $y = k$, then y becomes $+1$. If $y \neq k$, then y becomes -1. This partial bipolar regression produces coefficient candidates c_0^k and c_1^k, which are approximately in the general ballpark required for an initial guess.

Once all K discriminant formulas have been partially bipolar regressed, a single pass scoring run usually scores a CEP in the approximate range of from 5% to 20%, *in cases where the GPCM is an exact match for the dependent variable Y*, even when K grows larger.

Obviously, in cases where the GPCM is a poor match for the dependent variable Y, the returned CEP will remain in the 80% to 99% range as if the coefficients had been randomly chosen. This initial guess CEP discrepancy been exact matches and poor matches presents an evolutionary activity distribution opportunity which PBR measures by computing the PBR success rate. The PBR success rate is the percent inverse of CEP as follows:

$$\text{PBRSuccessRate} = 100\% - \text{CEP}. \tag{11.11}$$

The PBR success rate will be used in the next layer of the MDC algorithm to distribute evolutionary activity more efficiently.

11.2.2 Modified Sequential Minimization

The second layer of the MDC algorithm is an opportunistic modification of Platt's sequential minimization optimization algorithm, often used to train support vector machines [18].

At the start of the modified sequential minimization (MSM) layer, the candidate GPCM contains a swarm pool of a single set of coefficient constants which was produced by the PBR layer. Also the CEP for this single entry in the swarm pool and the PBR success rate are both available. The MSM layer multiplies the PBR success rate times 100 to produce the MSM repetition count, which determines the number of times the MSM will repeat without CEP improvement. The higher the PBR success rate, the greater the count of times the MSM will repeat. So the PBR success rate determines the evolutionary activity spent on the current candidate. Better candidates receive more evolutionary activity. Worse candidates receive less evolutionary activity.

For each repetition of the MSM, on the current candidate GPCM, the most fit entry in the swarm pool is chosen and a single erroneous training point is chosen at random. Since the chosen training point is in error, we know that its estimated dependent variable will not match the actual dependent variable, i.e., $ey \neq y$. Let us assume that $ey = i$ and that $y = j$. We therefore know that two discriminant formulas have the following relationship:

$$(c_0^i + (c_1^i \times D^i(x))) > (c_0^j + (c_1^j \times D^j(x))). \tag{11.12}$$

And we also know that, for that single erroneous training point, the maximum of all discriminant formulas, other than i and j, which we shall name α, will be also be less than $c_0^i + (c_1^i \times D^i(x)))$, which value we shall name β. Furthermore, we know that we can convert this single erroneous training point into a successful match if we alter the i and j coefficients such that the following relationships becomes true:

$$\begin{aligned} \alpha \leq & (c_0^i + (c_1^i \times D^i(x))) < \\ & (c_0^j + (c_1^j \times D^j(x))) \leq \beta. \end{aligned} \tag{11.13}$$

We select a value between α and β at random then alter c_0^j and c_1^j such that $(c_0^j + (c_1^j \times D^j(x)))$ is equal to the randomly selected value, and then alter c_0^i and c_1^i such that $(c_0^i + (c_1^i \times D^i(x)))$ is slightly less than the selected value. Of course we do not know what havoc these alterations will create for the other training points, but we do know that this selected training point will be converted from an error to a match.

A single pass scoring run is performed using the newly altered coefficients. The resulting CEP and the altered coefficients are inserted into the swarm pool sorted by CEP. If the new CEP is an improvement then we repeat the MSM layer once again. If the new CEP is *not* an improvement, then we decrement the MSM repetition count.

If the MSM repetition count is greater than zero then we repeat the MSM layer once again; otherwise, we terminate the MSM layer.

Upon termination of the final MSM repetition the final CEP is used to compute the MSM success rate as follows:

$$\text{MSMSuccessRate} = 100\% - \text{CEP}. \tag{11.14}$$

The MSM success rate will be used in the next layer of the MDC algorithm to distribute evolutionary activity more efficiently.

11.2.3 Bees Swarm Optimization

The third layer of the MDC algorithm is the repeated application of the well known bees swarm optimization algorithm often used in swarm evolution [4].

At the start of the bees swarm optimization (BSO) layer, the candidate GPCM contains a swarm pool of several sets of coefficient constants which were produced by the MSM layer. Also the CEP for the most fit entry in the swarm pool and the MSM success rate are both available. The BSO layer multiplies the MSM success rate times 100 to produce the BSO repetition count, which determines the number of times the BSO will repeat without CEP improvement. The higher the MSM success rate, the greater the count of times the BSO will repeat. So the MSM success rate determines the evolutionary activity spent on the current candidate. Better candidates receive more evolutionary activity. Worse candidates receive less evolutionary activity.

For each repetition of the BSO, the standard bees optimization algorithm is applied [4]. Then a single pass scoring run is performed using the new coefficient alterations. The resulting CEP and the altered coefficients are inserted into the swarm pool sorted by CEP. If the new CEP is an improvement then we repeat the BSO layer once again. If the new CEP is NOT an improvement then we decrement the BSO repetition count. If the BSO repetition count is greater than zero then we repeat the BSO layer once again; otherwise, we terminate the BSO layer.

At the end of the BSO layer, the coefficients of the current GPCM candidate are fully optimized *as far as the MDC algorithm is concerned.*

11.3 Computational Effort Distribution

In this section we create the several theoretical test problems which demonstrate the manner in which the MDC algorithm distributes computational effort as it attempts to optimize the coefficients of a candidate GPCM. All of these theoretical test problems attempt to optimize coefficients for a single artificially created GPCM against a target which has been explicitly created to demonstrate test cases which are poorly matched versus test cases which are well matched. Using the following formula (11.15), we will create a theoretical test data set using the following five discriminant formulas. Therefore we are trying to classify across $K = 5$ classes:

$$\begin{aligned} y = nominal(&3.4 + (1.57 \times x_0), \\ &2.1 - (-39.34 \times (x_4 \times x_1)), \\ &(2.13 \times (x_2 \div x_3)), \\ &1.0 - (46.59 \times (x_3 \times x_7)), \\ &(11.54 \times x_4)). \end{aligned} \tag{11.15}$$

Next we will use the MDC algorithm to optimize coefficients for the following candidate GPCMs, each with five proposed discriminants:

C01 $nomial(x_0, x_1, x_2, x_3, x_4)$
C02 $nomial(x_0, (x_1 \times x_4), x_2, x_3, x_4)$
C03 $nomial(x_0, (x_1 \times x_4), (x_2 \div x_3), x_3, x_4)$
C04 $nomial(x_0, (x_1 \times x_4), (x_2 \div x_3), (x_3 \times x_7), x_4)$.

The results of using the MDC algorithm to optimize each of the four test GPCMs, on the theoretical test data created with formula (11.15) are shown in Table 11.1. Notice how the fitness Classification Error Percent (CEP) improves as the tests move closer to matching formula (11.15). Also notice how the computational effort increases as the tests move closer to matching formula (11.15). The MDC algorithm distributes more computational effort to the more promising candidates and less computational effort to the less promising candidates.

Deterministic algorithms, like M2GP, apply the same amount of computational effort toward optimizing the coefficients of each GPCM equally. Whereas the MDC algorithm distributes the computational effort unevenly throughout the entire evolutionary process with less promising candidates receiving less computational effort and more promising candidates receiving more computational effort.

Table 11.1 MDC evolutionary effort

Test	PBR	MSM	BSO	CEP
C01	6	49	48	0.4550
C02	6	73	121	0.2187
C03	6	119	235	0.2033
C04	6	229	511	0.0000

PBR: number of PBR layer attempts; *MSM*: number of MSM layer attempts; *BSO*: number of BSO layer attempts; *CEP* fitness score of the optimized GPCM on test data

11.4 Theoretical Test Problems

The MDC algorithm is NOT adequate for performing a whole symbolic multi-class classification run. MDC is only adequate to optimize the coefficients of a selected GPCM candidate. Therefore we will be required to develop a temporary symbolic multi-class classification strategy wrapped around the MDC algorithm in order to proceed with testing.

Since this is a preliminary paper in a planned series of papers investigating extreme accuracy algorithms *vis vis* symbolic multi-class classification, we will create a first draft multi-class classification strategy as a temporary methodology to allow further testing. We do not yet know whether MDC is the most propitious coefficient optimization algorithm, or whether some hybrid of M2GP and MDC, or some other algorithm might be better. We are just getting started in our investigation.

Our temporary symbolic multi-class classification algorithm, to wrap around the MDC algorithm, will be composed of K+1 separate search islands—a general search island and a specific search island for each class. During the classification run, all $K + 1$ search islands will use Pareto front optimization to select GPCM candidates, optimize their discriminant coefficients with MDC, then score the final CEP. Whenever a new global most fit GPCM is discovered—in any of the $K + 1$ search islands—the K specialized search islands are reset to the new global most fit GPCM. Each k-th specialized search island then attempts to further optimize the best GPCM by holding all discriminant functions fixed, *except* the k-th discriminant function which is evolved using Pareto front evolution.

Our temporary Pareto front search strategy surrounds the MDC algorithm with a general attempt to find the best GPCM (*search island 0*) and K attempts specific to each of the K classes (*search islands 1 through* K) to find the best GPCM. No assertion is made that this temporary Pareto front strategy will be the best or final search strategy achieving extreme accuracy in symbolic multi-class classification. It is just a temporary search strategy to allow us to proceed with testing.

In this section we create four theoretical training and separate test data sets using discriminant formulas (T1) thru (T4). Each theoretical test problem has $K = 5$ classes. Tables 11.2 and 11.3 show the results for each of the four symbolic classification runs.

Table 11.2 Theoretical test problems

Name	Detail
T01	$y = nomial(1.57 \times x_0, -39.34 \times x_1, 2.13 \times x_2, 46.59 \times x_3, 11.54 \times x_4)$
T02	$y = nomial(3.4 + (1.57 \times x_0), 2.1 - (-39.34 \times (x_4 \times x_1)), 2.13 \times (x_2 \div x_3), 1.0 - (46.59 \times (x_5 \times x_5)), 11.54 \times x_4)$
T03	$y = nomial(3.4 + (1.57 \times x_0), 2.1 + (-39.34 \times min((x_4 \times x_1), x_6)), 2.13 \times ((x_2 \div x_3) \leq (x7)), 1.0 - (46.59 \times (x_5 \times x_9)), 11.54 \times x_4)$
T04	$y = nomial(3.4 + (1.57 \times (x_0 \leq x_1 1)), 2.1 + (9.34 \times min(x_4 \times x_1, x_6)), 2.13 \times max((x_2 \div x_3) \leq x_7, x_1 9), 1.0 - (46.59 \times max(x_1 5, x_8)), 11.54 \times x_4)$

Table 11.3 MDC theoretical test problems results

Name	WFFs	Train-hrs	Train-CEP	Test-CEP
T01	5K	0.59	0.0000	0.0000
T02	10K	2.011	0.0398	0.0411
T03	71K	10.08	0.0606	0.0810
T04	76K	10.08	0.0964	0.0976

WFFs (Well-formed formulas): number of regression candidates tested before finding a solution; *Train-hrs*: elapsed hours spent training on the training data; *Train-CEP*: fitness score of the champion on the noiseless training data; *Test-CEP*: fitness score of the champion on the noiseless testing data; this column has 0.0880 average fitness

The theoretical testing demonstrates the computational ease of using the MDC algorithm for GPCM coefficient optimization. Training is quick and grows only linearly as the number of classes and training points grow larger. The first simple test (*T01*) achieves extreme accuracy because the temporary Pareto front strategy selected an exact match GPCM during the run. As one would expect the CEP fitness scores get worse as the target discriminant functions grow more mathematically complex—*further from linear*. This is the result of the temporary Pareto front strategy not selecting exact match GPCMs during the run. A problem which will have to be addressed in the future papers if we are to achieve extreme accuracy in symbolic multi-class classification.

11.5 Real Data Test Problems

In this section we apply the MDC algorithm wrapped in its temporary Pareto front strategy on real world test problems taken from several sources. Some test data sets were downloaded from the University of California at Irvine machine learning repository (https://archive.ics.uci.edu/ml/datasets.html); other test data sets were downloaded from the Broad Institute cancer data sets (at http://www.broadinstitute.org/cgi-bin/cancer/datasets.cgi).

Another Volatility data set was constructed from the Yahoo downloadable VIX and UVXY daily historical data sets. This test problem attempted to classify the next day's profit or loss in the UVXY ETF, entirely from the previous day's percent change in the VIX and the percent change in the 140 day moving average of the VIX (Table 11.4).

The results of running the MDC algorithm wrapped in its temporary Pareto front strategy on these real world test problems is shown in Table 11.5.

This early in our planned investigation of extreme accuracy in multi-class classification, we did not expect to achieve the results shown in Table 11.5. While we are very far from any industrially usable techniques, the MDC algorithm wrapped in its temporary Pareto front strategy achieved a perfect CEP score on the Broad

Table 11.4 Real data test problems

Name	Decription
R01	Acute Myeloid Leukemia (Broad Institute)
R02	Iris (UCI)
R03	Heart disease (UCI)
R04	Volatility (Yahoo! VIX and UVXY historical data)
R05	Bank marketing (UCI)

Table 11.5 MDC real data test problems results

Test	WFFs	Train-hrs	Train-CEP	Test-CEP
R01	9K	0.24	0.0277	0.0000
R02	10K	0.11	0.0133	0.0134
R03	548K	10.00	0.0912	0.0925
R04	9K	0.14	0.0704	0.0705
R05	9K	0.29	0.1077	0.1097

WFFs (Well-formed formulas): number of regression candidates tested before finding a solution; *Train-hrs*: elapsed hours spent training on the training data; *Train-CEP*: fitness score of the champion on the noiseless training data; *Test-CEP*: fitness score of the champion on the noiseless testing data, with 0.0681 average fitness

Institute's leukemia data. The UCI Iris data also got a very good CEP score. All other real world test data sets got reasonable CEP scores. The Volatility test data achieved a reasonable CEP score and categorized, without any losses, the day's when the UVXY next day profit was 20% or above. There were four estimated trading signals all of which resulted in next day UVXY profits of 20% or more.

11.6 Conclusion

In a previous series of papers [5–10], significant accuracy issues were identified for state of the art symbolic regression systems and a comprehensive multi-island strategy for achieving extreme accuracy on a large well defined set of theoretical problems was developed and tested both in noiseless and in noisy training environments. Unfortunately these SR techniques do not translate directly into an algorithm which will achieve extreme accuracy on symbolic multi-class classification problems.

A first step in a planned investigation of extreme accuracy in symbolic multi-class classification is taken with the introduction of the Multilayer Discriminant Classification algorithm for optimizing the discriminant coefficients in a multi-class discriminant equation. Distribution pof computational effort Tests on the MDC algorithm demonstrate the desired properties of a high level of accuracy for exact match GPCMs and also well behaved distribution of computational effort with more promising GPCM candidates receiving more computational effort and less promising GPCM candidates receiving less computational effort.

Also, both theoretical and real world testing of the MDC algorithm *wrapped in a temporary Pareto front strategy* resulted in preliminary but promising CEP scores. It is now a reasonable suspicion that the same symbolic regression accuracy issues, *due primarily to the poor surface conditions of specific subsets of the problem space*, are also present in and obstructing extreme accuracy in symbolic multi-class classification problems.

Future research must explore the possibility of developing an Extreme Accuracy algorithm for the field of symbolic multi-class classification. Furthermore any such extreme accuracy algorithm would ideally be accompanied by a formal or informal proof of extreme accuracy on a well defined set of theoretical problems.

Finally, to the extent that the reasoning in even an *informal argument* of extreme accuracy can gain academic and commercial acceptance, a climate of *belief* in symbolic multi-class classification can be created wherein SC is increasingly seen as a "must have" tool in the scientific arsenal.

Truly knowing the strengths and weaknesses of our tools is an essential step in gaining trust in their use.

References

1. Castelli, M., Silva, S., Vanneschi, L., Cabral, A., Vasconcelos, M.J., Catarino, L., Carreiras, J.M.B.: Land cover/land use multiclass classification using gp with geometric semantic operators. In: Esparcia-Alcazar, A.I., Cioppa, A.D., De Falco, I., Tarantino, E., Cotta, C., Schaefer, R., Diwold, K., Glette, K., Tettamanzi, A., Agapitos, A., Burrelli, P., Merelo, J.J., Cagnoni, S., Zhang, M., Urquhart, N., Sim, K., Ekart, A., Fernandez de Vega, F., Silva, S., Haasdijk, E., Eiben, G., Simoes, A., Rohlfshagen, P. (eds.) Applications of Evolutionary Computing, EvoApplications 2013: EvoCOMNET, EvoCOMPLEX, EvoENERGY, EvoFIN, EvoGAMES, EvoIASP, EvoINDUSTRY, EvoNUM, EvoPAR, EvoRISK, EvoROBOT, EvoSTOC. Lecture Notes in Computer Sscienc, vol. 7835, pp. 334–343. Springer, Vienna (2013). https://doi.org/10.1007/978-3-642-37192-9_34
2. Gandomi, A.H., Alavi, A.H., Ryan, C. (eds.): Handbook of Genetic Programming Applications. Springer, Berlin (2015). https://doi.org/10.1007/978-3-319-20883-1
3. Ingalalli, V., Silva, S., Castelli, M., Vanneschi, L.: A multi-dimensional genetic programming approach for multi-class classification problems. In: Nicolau, M., Krawiec, K., Heywood, M.I., Castelli, M., Garcia-Sanchez, P., Merelo, J.J., Rivas Santos, V.M., Sim, K. (eds.) 17th European Conference on Genetic Programming. Lecture Notes in Computer Science, vol. 8599, pp. 48–60. Springer, Granada (2014). https://doi.org/10.1007/978-3-662-44303-3_5
4. Karaboga, D., Akay, B.: A survey: algorithms simulating bee swarm intelligence. Artif. Intell. Rev. **31**(1–4), 61–85 (2009)
5. Korns, M.F.: Abstract expression grammar symbolic regression. In: Riolo, R., McConaghy, T., Vladislavleva, E. (eds.) Genetic Programming Theory and Practice VIII. Genetic and Evolutionary Computation, vol. 8, chap. 7, pp. 109–128. Springer, Ann Arbor (2010). http://www.springer.com/computer/ai/book/978-1-4419-7746-5
6. Korns, M.F.: Accuracy in symbolic regression. In: Riolo, R., Vladislavleva, E., Moore, J.H. (eds.) Genetic Programming Theory and Practice IX, Genetic and Evolutionary Computation, chap. 8, pp. 129–151. Springer, Ann Arbor (2011). https://doi.org/10.1007/978-1-4614-1770-5_8

7. Korns, M.F.: A baseline symbolic regression algorithm. In: R. Riolo, E. Vladislavleva, M.D. Ritchie, J.H. Moore (eds.) Genetic Programming Theory and Practice X, Genetic and Evolutionary Computation, chap. 9, pp. 117–137. Springer, Ann Arbor (2012). https://doi.org/10.1007/978-1-4614-6846-2_9
8. Korns, M.F.: Extreme accuracy in symbolic regression. In: Riolo, R., Moore, J.H., Kotanchek, M. (eds.) Genetic Programming Theory and Practice XI, Genetic and Evolutionary Computation, chap. 1, pp. 1–30. Springer, Ann Arbor (2013). https://doi.org/10.1007/978-1-4939-0375-7_1
9. Korns, M.F.: Extremely accurate symbolic regression for large feature problems. In: Riolo, R., Worzel, W.P., Kotanchek, M. (eds.) Genetic Programming Theory and Practice XII, Genetic and Evolutionary Computation, pp. 109–131. Springer, Ann Arbor (2014). https://doi.org/10.1007/978-3-319-16030-6_7
10. Korns, M.: Highly accurate symbolic regression with noisy training data. In: Riolo, R., Worzel, W.P., Kotanchek, M., Kordon, A. (eds.) Genetic Programming Theory and Practice XIII, Genetic and Evolutionary Computation. Springer, Ann Arbor (2015). https://doi.org/10.1007/978-3-319-34223-8. http://www.springer.com/us/book/9783319342214
11. Kotanchek, M., Smits, G., Vladislavleva, E.: Trustable symbolic regression models: using ensembles, interval arithmetic and pareto fronts to develop robust and trust-aware models. In: Riolo, R.L., Soule, T., Worzel, B. (eds.) Genetic Programming Theory and Practice V, Genetic and Evolutionary Computation, chap. 12, pp. 201–220. Springer, Ann Arbor (2007). https://doi.org/10.1007/978-0-387-76308-8_12. http://citeseerx.ist.psu.edu/viewdoc/summary?doi=10.1.1.457.5272
12. Koza, J.R.: Genetic Programming: On the Programming of Computers by Means of Natural Selection. MIT Press, Cambridge, MA (1992). http://mitpress.mit.edu/books/genetic-programming
13. Koza, J.R.: Genetic Programming II: Automatic Discovery of Reusable Programs. MIT Press, Cambridge, MA (1994). http://www.genetic-programming.org/gpbook2toc.html
14. Koza, J.R., Andre, D., Bennett III, F.H., Keane, M.: Genetic Programming III: Darwinian Invention and Problem Solving. Morgan Kaufman (1999). http://www.genetic-programming.org/gpbook3toc.html
15. Langdon, W.B., Poli, R.: Foundations of Genetic Programming. Springer, Berlin (2002). https://doi.org/10.1007/978-3-662-04726-2. http://www.cs.ucl.ac.uk/staff/W.Langdon/FOGP/
16. McConaghy, T.: Ffx: Fast, scalable, deterministic symbolic regression technology. In: Riolo, R., Vladislavleva, E., Moore, J.H. (eds.) Genetic Programming Theory and Practice IX, Genetic and Evolutionary Computation, chap. 13, pp. 235–260. Springer, Ann Arbor (2011). https://doi.org/10.1007/978-1-4614-1770-5_13. http://trent.st/content/2011-GPTP-FFX-paper.pdf
17. Nelder, J., Wedderburn, R.: Generalized linear models. Stat. Soc **135**, 370–383
18. Platt, J.: Sequential minimal optimization: A fast algorithm for training support vector machines. Technical Report Microsoft Research Technical Report MSR-TR-98-14 (1998)
19. Poli, R., McPhee, N.F., Vanneschi, L.: Analysis of the effects of elitism on bloat in linear and tree-based genetic programming. In: Riolo, R.L., Soule, T., Worzel, B. (eds.) Genetic Programming Theory and Practice VI, Genetic and Evolutionary Computation, chap. 7, pp. 91–111. Springer, Ann Arbor (2008). https://doi.org/10.1007/978-0-387-87623-8_7
20. Smits, G., Kotanchek, M.: Pareto-front exploitation in symbolic regression. In: O'Reilly, U.M., Yu, T., Riolo, R.L., Worzel, B. (eds.) Genetic Programming Theory and Practice II, chap. 17, pp. 283–299. Springer, Ann Arbor (2004). https://doi.org/10.1007/0-387-23254-0_17

Chapter 12
A Generic Framework for Building Dispersion Operators in the Semantic Space

Luiz Otavio V. B. Oliveira, Fernando E. B. Otero, and Gisele L. Pappa

Abstract This chapter proposes a generic framework to build geometric dispersion (GD) operators for Geometric Semantic Genetic Programming in the context of symbolic regression, followed by two concrete instantiations of the framework: a multiplicative geometric dispersion operator and an additive geometric dispersion operator. These operators move individuals in the semantic space in order to balance the population around the target output in each dimension, with the objective of expanding the convex hull defined by the population to include the desired output vector. An experimental analysis was conducted in a testbed composed of sixteen datasets showing that dispersion operators can improve GSGP search and that the multiplicative version of the operator is overall better than the additive version.

Keywords Genetic programming · Geometric semantic genetic programming · Dispersion operators · Behavioral diversity · Symbolic regression · Search operators

12.1 Introduction

The role of the crossover operator in tree-based genetic programming has been a discussion point for a long time [2], as many researchers believed the lack of context associated with the tree nodes makes crossover to resemble a macro mutation. In semantic genetic programming algorithms, in particular their geometric counterparts, this is mitigated by making syntactic modifications more semantically-aware—i.e., focusing on how syntactic modifications reflect on the semantics of the individuals.

L. O. V. B. Oliveira · G. L. Pappa (✉)
DCC, Universidade Federal de Minas Gerais, Belo Horizonte, Brazil

F. E. B. Otero
School of Computing, University of Kent, Chatham Maritime, UK

R. Riolo et al. (eds.), *Genetic Programming Theory and Practice XIV*, Genetic and Evolutionary Computation,
https://doi.org/10.1007/978-3-319-97088-2_12

This chapter deals with the problem of symbolic regression, where the semantics of an individual is defined as a point in a n-dimensional space, called semantic space, and n is the number of examples in the training set. In geometric semantic genetic programming (GSGP), the geometric semantic crossover and mutation operators [13] guarantee that the semantic fitness landscape explored by the GP is conic, which has a positive impact in the search process. The problem is then how long GSGP might take to find the optimum.

The challenge of finding the optimal solution or not is then dependent on other components of GSGP. For example, as the GSGP crossover operator produces offspring by performing a convex combination of its parents, the set of candidate individuals generated during evolution is delimited by the *convex hull*[1] of the semantics of the current population [16]. Hence, if the target output is not within the convex hull, the algorithm will never be able to find it using crossover alone. The mutation operator deals with this problem by expanding the convex hull. However, GSGP might take a prohibitive amount of time to get to the relevant regions of the search space depending on the distribution of the individuals in the initial generation.

In this context, Oliveira et al. [15] presented a heuristic operator to move individuals through the semantic space in order to, hopefully, include the target output inside the convex hull defined by the current population. The operator, called geometric dispersion (GD), applies multiplicative constants to the individuals aiming to balance the proportion of the population on the left and right side of the target output in each dimension of the semantic space.

In this same direction, this chapter proposes a generic framework for geometric dispersion operators allowing different mathematical operations to redistribute the population. The operation used to add the constant to the individual has direct impact on the way it is moved through the space. Thus, other operations, besides the multiplication used in the original GD, allow the resulting individual to reach other regions of the semantic space with different effects on the search. The framework is used to build a geometric dispersion operator based on the addition operation and evaluates the impact of the new operator on the evolution. We performed an experimental analysis in a test bed composed of sixteen datasets. We compared the results obtained by GSGP with the multiplicative and the additive versions of the geometric dispersion, tested separately, and with the GSGP without the dispersion operators. Results indicate dispersion operators have a positive impact on the search, improving the root mean square error in relation to the GSGP without this operator.

The remaining of this chapter is organised as follows. Section 12.2 provides an overview to GSGP for symbolic regression problems and the crossover limitation regarding the population's convex hull. Section 12.3 reviews previous works involving the convex hull described by the population in GSGP and Sect. 12.4 presents a framework for GD operators along with two particular implementations.

[1]The *convex hull* of a set of points is given by the set of all possible convex combinations of these points [18].

Section 12.5 presents the experimental analysis in sixteen different datasets followed by conclusions and research directions in Sect. 12.6.

12.2 Background

Most genetic programming algorithms employ *traditional* genetic operators that perform syntactic modification on individuals in order to change their behaviour—the behaviour of an individual is referred to as its semantics. One particular drawback of traditional genetic operators is that there is no guarantee that syntactic modifications will lead to different behaviour. Therefore, they represent an indirect way of changing the semantics of an individual. Geometric semantic genetic programming (GSGP) [13], on the other hand, employ semantic genetic operators to introduce syntactic modification on individuals that guarantee to change their semantics.

In this chapter, we focus on GSGP applied to symbolic regression problems. Symbolic regression problems can be seen as a supervised learning procedure: given a finite set of input-output pairs representing the fitness cases, defined as $T = \{(\mathbf{x}_i, y_i)\}_{i=1}^{n}$—where $(\mathbf{x}_i, y_i) \in \mathbb{R}^d \times \mathbb{R}$ $(i = 1, 2, \ldots, n)$—symbolic regression consists in inducing a model $p : \mathbb{R}^d \rightarrow \mathbb{R}$ that maps inputs to outputs, such that $\forall(\mathbf{x}_i, y_i) \in T : p(\mathbf{x}_i) = y_i$.

Let $I = [\mathbf{x}_1, \mathbf{x}_2, \ldots, \mathbf{x}_n]$ and $O = [y_1, y_2, \ldots, y_n]$ be the input and the output vectors,[2] respectively, associated to the fitness cases. The semantics of a program p represented by an individual evolved by GSGP, denoted as $s(p)$, is the vector of outputs it produces when applied to the set of inputs I, i.e., $s(p) = p(I) = [p(\mathbf{x}_1), p(\mathbf{x}_2), \ldots, p(\mathbf{x}_n)]$. This notation is extended to the semantics of a population of programs $P = \{p_1, p_2, \ldots, p_k\}$, i.e., $s(P) = \{s(p_1), s(p_2), \ldots, s(p_k)\}$. The semantics of any program can be represented as a point in a n-dimensional space S, referred to as the semantic space, where n is the number of fitness cases. Note that the desired output vector O can also be represented in the semantic space.

GSGP employs semantic geometric operators to evolve the individuals in a population. Let P' be the solution set comprising all the possible candidate solutions to a problem in the real domain, the geometric semantic crossover and mutation operators are defined as follows:

Definition 12.1 Given two parent programs $p_1, p_2 \in P'$, the geometric semantic crossover for the space of real functions $GSX : P' \times P' \rightarrow P'$ returns the real function

$$p_3 = r \times p_1 + (1 - r) \times p_2, \tag{12.1}$$

[2] Note that when $\mathbf{x}_i \in \mathbb{R}^d$ with $d > 1$ the vector I becomes a matrix with dimensions $d \times n$. We allow an abuse of notation by representing the matrix as a vector with dimension n, where each element corresponds to a vector of dimension d.

where r is a random real constant in [0, 1] (for fitness function based on Euclidean distance) or a random real function with codomain [0, 1] (for fitness function based on Manhattan distance).

Definition 12.2 Given a parent program $p \in P'$, the geometric semantic mutation for the space of real functions $GSM : P' \times \mathbb{R}^+ \rightarrow P'$ with mutation step ε returns the real function

$$p' = p + \varepsilon \times (r_1 - r_2), \tag{12.2}$$

where r_1 and r_2 are random real functions.

An interesting characteristic of GSGP is that the fitness of an individual p is the distance of its output vector $s(p)$ to the desired output vector O. Therefore, the fitness landscape induced by semantic genetic operators is unimodal by construction [13]. Despite the unimodal fitness landscape, the stochastic nature of these operators—as a result of using random real functions and constants—has been shown to be a more suitable way to explore the space in terms of generalisation, when compared to modifications of these operators where decisions are based on fitness cases error [1, 7, 10]. The area defined by the set of individuals (points in the semantic space) define the convex hull of the population:

Definition 12.3 The *convex hull* of a set H of points in $\mathbb{R}^n$, denoted as $C(H)$, is the set of all convex combinations of points in H [18].

Let P be a population of individuals, we adopt the notation $C(s(P))$ to denote the convex hull of the set composed by the semantics of the individuals of P, i.e., $s(P)$. Since GSX is, by definition, a geometric crossover operator [13], we have the following theorem regarding the convex hull of the population:

Theorem 12.1 *Let P_g be the population at generation g. For a GSGP, where the GSX operator is the only search operator available, we have $C(s(P_{g+1})) \subseteq C(s(P_g)) \subseteq \ldots \subseteq C(s(P_1)) \subseteq C(s(P_0))$.*

Theorem 12.1 is a particular case of the Theorem 3 defined and proved by [12], and it has an important implication regarding the GSX operator. Given a population P and a semantic vector q in S, the offspring resulting from the application of GSX to any pair of individuals in P can reach q if and only if $q \in C(s(P))$. Consequently, if GSGP has no other search operators (only GSX), a semantic vector q is reachable only if $q \in C(s(P_0))$—i.e., if q is located inside the convex hull of the initial population.

Figure 12.1 illustrates this situation for a two-dimensional semantic space. Without loss of generality let $O = [0, 0]$ be the desired output vector defined by the training cases. Now consider two different populations P_a and P_b, where the individuals from P_a are concentrated in the upper-right side of O and, consequently, $C(s(P_a))$ cannot reach the origin O. On the other hand, the set $s(P_b)$ is distributed around the desired output such that $O \in C(s(P_b))$. In the first scenario, GSGP needs a mutation operator to expand the convex hull to reach O. In the second scenario, the

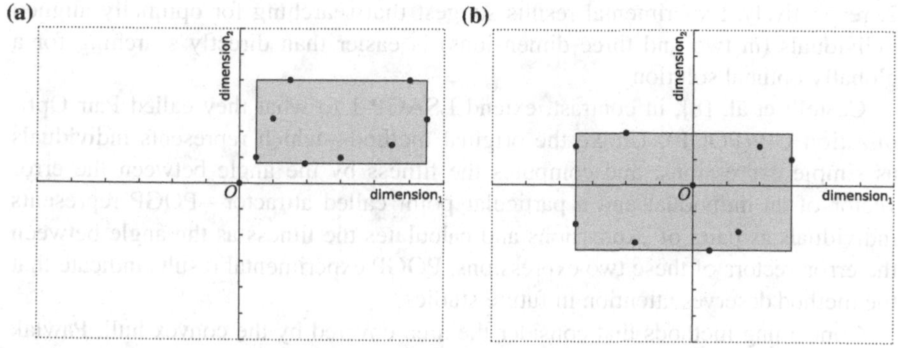

Fig. 12.1 Example of different distributions of a population in a two-dimensional semantic space. The desired output O is located in the origin of the space and the shaded area corresponds to the convex hull under the Manhattan distance. **(a)** P_a: population concentrated into a single quadrant. **(b)** P_b: population encompasses solutions in all quadrants

desired vector O can be reached using a crossover operator alone, as it is already inside the convex hull, or it can be calculated analytically[3] with no need to use GSGP.

12.3 Related Work

Previous work on GSGP have proposed different approaches to take advantage of the properties of the geometric semantic space to improve search. However, to the best of our knowledge, so far only two have investigated ways to increase the area covered by the convex hull of the population—in particular focusing on the coverage of the initial population, as discussed in this section.

Regarding operators that take advantage of the conic shape of the geometric semantic space, Ruberto et al. [19] explore the geometry of the semantic space through the concept of error vector. An error vector is represented by a point in the n-dimensional space, called error space, given by the translation $t_e(p) = s(p) - O$. This notion is used to introduce the concept of optimally aligned individuals in the error space, i.e., given a number of dimensions $\mu = 2, 3, \ldots, n$, where n is the size of the training set, μ individuals are optimally aligned in the error space if they belong to the same μ-dimensional hyperplane intersecting the origin of the error space. The authors show that if μ individuals are optimally aligned, we can analytically obtain an equation to express the target output vector O. In this context, they present GP-based methods to find optimally aligned individuals in two and three dimensions, called ESAGP-1 (Error Space Alignment GP) and ESAGP-

[3]The coefficients of convex combinations can be found by means of Gaussian elimination [9].

2, respectively. Experimental results suggest that searching for optimally aligned individuals (in two and three dimensions) is easier than directly searching for a globally optimal solution.

Castelli et al. [8], in contrast, extend ESAGP-1 to what they called Pair Optimization GP (POGP). Unlike the original method—which represents individuals as simple expressions, and computes the fitness by the angle between the error vector of an individual and a particular point called attractor—POGP represents individuals as pairs of expressions and calculates the fitness as the angle between the error vectors of these two expressions. POGP experimental results indicate that the method deserves attention in future studies.

Concerning methods that consider the area covered by the convex hull, Pawlak [16] proposed the Competent Initialization (CI) method, which aims to increase the convex hull of the initial population. The algorithm adopts a generalized version of the Semantically Driven Initialization (SDI) method [3], initially proposed for non-geometric spaces, to generate individuals semantically distinct. SDI randomly picks a node from the function set to combine individuals already in the population. If the resulting program has semantics different from other individuals of the population, it is accepted; otherwise, the method makes a new attempt of generating an individual. The process continues until a semantically distinct individual is created, following a trial-and-error strategy. CI, on the other hand, accepts the semantically distinct individual only if it is not in the current convex hull. The main drawback of both SDI and CI methods is the possible waste of resources, since individuals are randomly created, evaluated and discarded when they are semantically similar to an existing individual of the population or when it is already in the population's convex hull.

The Semantic Geometric Initialization (SGI) [17], on the other hand, generates a set S of semantics, such that the desired output is guaranteed to belong to the convex hull of S. These semantics are generated by adding or subtracting an offset to O in different combinations of the semantic space dimensions. Then, for each semantics $s_i \in S$, the method generates an individual whose semantics is equal to s_i. The synthesis of these individuals is domain dependent and authors presented methods to generate individuals for symbolic regression domain (by polynomial interpolation) and for boolean domain (by a boolean formula). The experimental analysis indicates that SGI can achieve training error significantly smaller than the ramped half-and-half method in symbolic regression and boolean problems. However, the test error achieved by SGI is significantly higher than the error achieved by ramped half-and-half [11], which indicates that SGI is very susceptible to overfitting.

Although not taking advantage of the geometric properties of the search space, Castelli et al. [5] proposed a semantic-based algorithm that keeps a distribution of different semantics during the evolution to drive GP to search in areas of the semantic space where previous good solutions were found. The method outperformed standard GP and bacterial GP [4] in the test bed adopted. However, the individuals generated presented statistically bigger sizes than the individuals generated by the other two GP variants.

12.4 Geometric Dispersion Operators

In this section we present a generic geometric dispersion (GD) framework along with two implementations, the multiplicative geometric dispersion (MGD) [15] and the additive geometric dispersion (AGD) operators.

12.4.1 A Framework for Geometric Dispersion Operators

This section presents a general framework for geometric dispersion (GD)[4] operators aiming to redistribute the population around the desired output vector O in the semantic space. These operators move a given individual to the region of the semantic space around O with the lowest concentration of individuals in order to, hopefully, modify the convex hull of the population to contain the desired output.

GD operators adopt a greedy strategy to redistribute the population around O by examining each dimension of S separately. For each dimension of the semantic space, GD computes the proportion of individuals whose semantics is greater than and less than O for that dimension. The method uses this information to move the individuals through the semantic space—by means of mathematical operations applied to the individual's program—in order to balance each one of the dimensions of S.

When we know the region of the semantic space around O where we want to have individuals shifted to, different methods can be used to move individual p. GD operators do that by applying a constant m to p through a mathematical operation $\oplus$, in the form $m \oplus p$. The movement performed by the GD operator depends directly of the chosen operation for $\oplus$. Thus, the value of m must be chosen such that the displacement of p benefits the largest number of dimensions.

The process of finding this value is equivalent to find m that solves the inequality system:

$$\begin{cases} m \oplus s(p)[1] \lessgtr_1 O[1] \\ m \oplus s(p)[2] \lessgtr_2 O[2] \\ \dots \\ m \oplus s(p)[k] \lessgtr_k O[k] \end{cases} \qquad (12.3)$$

where '$\lessgtr$' is a inequality operator ('<' or '>') chosen according to the asymmetry of the population.

[4] Oliveira et al. [15] presents the first geometric dispersion operator. However, this operator is a particular case of the framework presented in this paper. Hence, hereafter their operator is referred as multiplicative geometric dispersion (MGD) operator in contrast to the GD framework.

Let GT (greater than) and LT (less than) be n-element arrays, where the i-th element corresponds to the number of individuals p_k in the current population P where $s(p_k)[i] > O[i]$ and $s(p_k)[i] < O[i]$, respectively. If $GT[i] > LT[i]$, the population is unbalanced with more individuals in the right side of the desired output vector in the i-th dimension of the semantic space, and the individual should be moved to the left side of O—the symbol '$\lesseqgtr_i$' is replaced by '<'. Otherwise, if $GT[i] < LT[i]$, the imbalance occurs in the opposite side, i.e., the population is concentrated on the left side of O in the dimension i and the individual should be moved to the opposite side of O—the symbol '$\lesseqgtr_i$' is replaced by '>'. Note that when $GT[i] = LT[i]$, no inequalities are added to the system. Therefore the number of inequalities in Eq. (12.3) is less or equal to the number of dimensions, i.e., $k \leq n$.

However, due to the large number of inequalities in the system, usually it does not admit feasible solutions. Thus, instead of finding a value for m that satisfies all inequalities, the operator finds one that maximizes the number of satisfied inequalities. Oliveira et al. [15] present algorithms to both construct the system of inequalities and find the value of m that satisfies the largest number of inequalities when the mathematical operation adopted is the multiplication, i.e., $\oplus$ is $\times$ (times). We generalise these algorithms and present a framework, called geometric dispersion (GD), which moves individuals through the semantic space in order to distribute the population around O.

GD is independent of the arithmetic operation adopted in the inequalities. The only requirements are that the operation is binary and allows inverse. Let $\oplus$ and $\ominus$ be a binary operation and its inverse, respectively. The variable m can be isolated in the left side of the system of inequalities of Eq. (12.3) as shown in Eq. (12.4), such that the operator can find the value that satisfies the largest number of inequalities.

$$\begin{cases} m \lesseqgtr_1 O[1] \ominus s(p)[1] \\ m \lesseqgtr_2 O[2] \ominus s(p)[2] \\ \dots \\ m \lesseqgtr_k O[k] \ominus s(p)[k] \end{cases} \tag{12.4}$$

Algorithm 1 introduces the procedure to define the system of inequalities. Given the arrays GT and LT, it checks each dimension i for an unbalanced distribution, i.e., where $GT[i] \neq LT[i]$. When these values differ, the method adds a new inequality to the system, represented by a bound value that should be satisfied.

The sign '$\lesseqgtr$' of the inequalities is defined according to the distribution of the population in the verified dimension (lines 4–8). If $GT[i] > LT[i]$, then '$\lesseqgtr_i$' is replaced by '<'. Otherwise, if $GT[i] < LT[i]$, it is replaced by '>'.

The next step of the method is to isolate m in the left side of the inequality and store the value of the right side in *bound* (lines 9–10). There are a few considerations in this step, according to the arithmetic operation used in the inequalities. E.g., if GD uses multiplication ($\oplus$ is $\times$), as presented by [15], the method must check for

Algorithm 1 GD procedure to build the system of inequalities

Require: Individual program (p), desired output (O), population distribution (GT, LT)

```
1: B ← {}
2: for i ← 1 to |s(p)| do                                    ▷ Calculate the bounds
3:     if GT[i] ≠ LT[i] then
4:         if GT[i] > LT[i] then
5:             inqSign ← 'lessThan'                         ▷ '≶_i' is replaced by '<'
6:         else
7:             inqSign ← 'greaterThan'                      ▷ '≶_i' is replaced by '>'
8:         end if
9:         Isolate m in the left side of the inequality     ▷ m ≶_i O[i] ⊖ s(p)[i]
10:        bound ← right side value                     ▷ bound ← O[i] ⊖ s(p)[i]
11:        if inqSign = 'lessThan' then
12:            Add bound to B as upper bound
13:        else
14:            Add bound to B as lower bound
15:        end if
16:    end if
17: end for
18: return B
```

division by zero and negative value on the left side of the inequality. When a division by zero is found, the algorithm ignores the inequality. When the left side is negative, both sides of the inequality are multiplied by -1, inverting the inequality sign.

The sign of the inequalities is used to define the type of bound (lines 11–15). If the sign is '$<$', the value of m should be smaller than *bound* (it is an upper bound). Otherwise, m should be greater than *bound* (it is a lower bound). The bounds and their types are used to compute the value of m in Algorithm 2.

Algorithm 2 follows the method presented in [15]. It first sorts B by value in ascending order. The auxiliary variables *maxSatisfied*, *index* and *cSatisfied* store the number of inequalities satisfied by the best bound for m found so far, its index and the number of inequalities satisfied by the bound examined in the current iteration, respectively. The method starts by considering the interval before the first bound, i.e., $(-\infty, B[1].value)$. If a value from this interval is picked for m, all the upper bounds are satisfied, i.e., $maxSatisfied = n_{ub}$ (line 2). It then iterates over B counting the number of upper and lower bounds satisfied by each interval until $(B[i], \infty)$ (lines 5–16). If the examined value corresponds to an upper bound, we decrement the *cSatisfied* counter, since the interval in the right side of the bound does not satisfy it. On the other hand, if the examined value corresponds to a lower bound, the right side interval satisfies the bound and *cSatisfied* is incremented.

After finding the best interval for m, the procedure assigns an actual value for m (lines 17–30). If the best interval corresponds to $(-\infty, B[1])$ or to $(B[|B|], \infty)$, m takes the output of *getLeftExtreme* and *getRightExtreme*, respectively. Otherwise, the method selects a random value in the interval $(B[index], B[index + 1])$.

Algorithms 3 and 4 present the procedures *getRightExtreme* and *getLeftExtreme*, respectively. The control variable *shiftOne* indicates if the methods should use the

Algorithm 2 Finds *m*

```
Require: Set of bounds for m (B), shift control variable (shiftOne)
n_ub ← number of 'ub's in B
maxSatisfied ← cSatisfied ← n_ub
index ← 0
Sort B by value in ascending order
for i ← 1 to |B| do                                  ▷ Find the best interval for m
    bound ← B[i]
    if bound is a lower bound then
        cSatisfied ← cSatisfied + 1
        if cSatisfied > maxSatisfied then
            maxSatisfied ← cSatisfied
            index ← i
        end if
    else
        cSatisfied ← cSatisfied − 1
    end if
end for
if index = 0 then                                    ▷ Calculate m
    if B is empty then                               ▷ No need to move p
        m ← 1
    else
        m ← getLeftExtreme(B, shifOne)
    end if
else
    if index = |B| then
        m ← getRightExtreme(B, shifOne)
    else
        δ ← B[index + 1].value − B[index].value
        m ← B[index].value + δ × rnd()               ▷ rnd() returns a random value in (0, 1)
    end if
end if
return m
```

same strategy adopt by [15], i.e., shift values in the extreme of the interval by one. Otherwise, the algorithms shift the values by a random value proportional to the closest interval defined in B.

The value of m returned by Algorithm 2 is then used to move individual p in the semantic space. GD is applied during the evolution at every generation, right before other genetic semantic operators (crossover and mutation). The probability of applying a GD operator—individual-wise—pgd, as proposed by [15], is given by:

$$pgd_g = pgd_0 \times exp\left(\frac{-\alpha \times g}{g_{max}}\right), \tag{12.5}$$

where pgd_0 is the base probability, α is the decay rate, g is current generation index and g_{max} is total number of generations. Equation (12.5) ensures the probability of applying the operator decays exponentially with the generations.

12.4.2 Multiplicative Geometric Dispersion

The geometric dispersion operator proposed in [15], here called multiplicative geometric dispersion (MGD), is an implementation of the GD framework where the constant m is multiplied by the semantics of the individual p. MGD manipulates inequality systems as given by Eq. (12.3)—in the form $m \times s(p)[i] \lesseqgtr_i O[i]$—and isolates m in the left side as presented by Eq. (12.4)—in the form $m \lesseqgtr_i O[i]/s(p)[i]$, where $\oplus$ and $\ominus$ are replaced by $\times$ and $\div$, respectively. The multiplicative operation applied to the individual p is geometrically equivalent to moving it through the line crossing both $s(p)$ and the origin of S.

As discussed above, when isolating m in the i-th dimension, MGD must consider two special cases. First, if $s(p)[i] = 0$, isolating m implies in division by zero and the operator ignores the inequality. Second, if $s(p)[i] < 0$, the inequality is multiplied by -1 and the inequality sign is inverted. For instance, let $m \times (-2) > 4$ be one of the inequalities. Thus, as $s(p)[i] = -2 < 0$, the inequality is multiplied by -1 before isolating m in the left side, leading to $m \times 2 < -4$.

12.4.3 Additive Geometric Dispersion

Besides the MGD, we present a geometric dispersion operator based on addition. The additive GD (AGD) moves a given individual p through the line $L = \{s(p) + t : t \in \mathbb{R}\}$, with $L \subset S$, in order to redistribute the population around O. The inequalities used by AGD are in the form $m + s(p)[i] \lesseqgtr O[i]$, which result in $m \lesseqgtr O[i] - s(p)[i]$, where i is the dimension analysed.

The use of different mathematical operations within the GD operators allows them to explore different regions of S. Figure 12.2 presents an example in a two-dimensional semantic space. In order to keep $dimension_1$ balanced and balance $dimension_2$, it is necessary to move p to the upper-left side of O. However, in this example, only AGD can reach this region of the space.

Algorithm 3 *getRightExtreme* procedure

Require: Set of bounds for m (B), control variable *shiftOne*

```
1: if shifOne=TRUE or |B| < 2 then
2:     return B[|B|] + 1
3: else
4:     δ ← B[|B|].value − B[|B| − 1].value
5:     return B[|B|] + δ × rnd()
6: end if
```

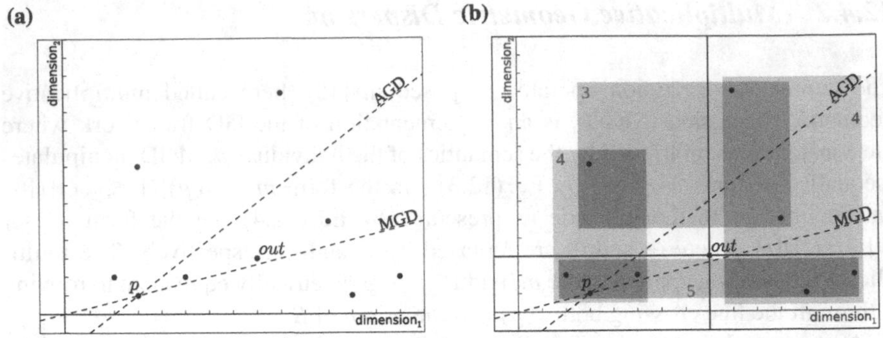

Fig. 12.2 Lines described by the AGD and MGD operators applied to an individual p. (**a**) AGD and MGD lines in the semantic space. (**b**) Distribution of individuals around O regarding $dimension_1$ (blue) and $dimension_2$ (red). The numbers indicate the frequency of individuals on each side of O for each dimension (Color figure online)

Algorithm 4 *getLeftExtreme* procedure

Require: Set of bounds for m (B), control variable *shiftOne*

```
1: if shifOne=TRUE or |B| < 2 then
2:     return B[1] − 1
3: else
4:     δ ← B[2].value − B[1].value
5:     return B[1] − δ × rnd()
6: end if
```

12.5 Experimental Analysis

This section presents an empirical analysis of the effect of different versions of the GD operator within GSGP. We compare the results obtained by GSGP with AGD (referred to as GSGP+A), GSGP with MGD [15] (referred to as GSGP+M) and GSGP without dispersion operators [6] in a test bed of sixteen symbolic regression datasets comprising both real-world and synthetic problems, as presented in Table 12.1. The test bed along with parameters adopted in the algorithms are the same from our previous work [15].

For each real-world dataset, we performed a 5-fold cross-validation with 10 replications, resulting in 50 executions. For the synthetic datasets (except *keijzer-6* and *keijzer-7*), we generated five different sets and, for each sample, applied the algorithms 10 times, resulting again in 50 executions. For keijzer-6 and keijzer-7, the test set is fixed, so we performed 50 executions. The categorical attributes, namely *vendor name* and *model name* from the cpu dataset and *month* and *day* from the forestFires dataset, were removed for compatibility purposes.

Table 12.1 Training RMSE (median and IQR) obtained by the algorithms for each dataset

Dataset	GSGP		GSGP+A		GSGP+AR		GSGP+M		GSGP+MR		GD param.	
	Med.	IQR	Med.	IQR	Med.	IQR	Med.	IQR	Med.	IQR	α	pgd_0
airfoil[a]	7.885	0.527	1.873	0.057	1.900	0.052	1.886	0.041	1.890	0.051	5	0.2
bioavailability[a]	9.893	0.652	9.653	0.552	9.706	0.547	9.695	0.690	9.794	0.703	10	0.4
concrete[a]	3.647	0.138	3.659	0.136	3.644	0.150	3.654	0.118	3.634	0.109	10	0.6
cpu[a]	6.126	0.665	6.149	0.977	6.223	1.067	6.151	0.905	6.215	0.790	10	0.4
energyCooling[a]	1.257	0.070	1.282	0.065	1.267	0.057	1.271	0.066	1.255	0.052	10	0.4
energyHeating[a]	0.802	0.113	0.790	0.084	0.789	0.123	0.798	0.083	0.764	0.084	10	0.2
forestfires[a]	30.737	4.626	31.684	4.858	31.247	4.490	30.967	4.201	31.534	4.156	10	0.4
keijzer-5[b]	0.045	0.003	0.063	0.006	0.062	0.006	0.026	0.008	0.026	0.009	0	0.6
keijzer-6[b]	0.007	0.005	0.007	0.007	0.007	0.005	0.007	0.006	0.006	0.005	0	0.2
keijzer-7[b]	0.017	0.010	0.017	0.010	0.016	0.009	0.016	0.009	0.014	0.009	5	0.2
ppb[a]	0.917	0.266	0.930	0.241	0.924	0.274	0.954	0.305	0.937	0.202	5	0.2
towerData[a]	20.436	0.610	20.558	0.704	20.587	0.610	20.405	0.621	20.472	0.404	10	0.2
vladislavleva-1[b]	0.012	0.002	0.012	0.002	0.012	0.001	0.012	0.002	0.012	0.002	10	0.6
vladislavleva-4[b]	0.038	0.001	0.038	0.002	0.038	0.002	0.038	0.001	0.038	0.002	10	0.2
wineRed[a]	0.493	0.011	0.494	0.012	0.494	0.010	0.494	0.011	0.495	0.010	10	0.2
wineWhite[a]	0.641	0.003	0.642	0.004	0.642	0.004	0.642	0.003	0.641	0.003	10	0.2

The last two columns present the parameters used as input in GD operators

[a] Real-world dataset

[b] Synthetic dataset

All executions used a population of 1000 individuals evolved for 2000 generations with tournament selection of size 10. The same random seed is employed to initialize the pseudorandom number generator in all methods. The grow method [11] was adopted to generate the random functions inside the geometric semantic crossover and mutation operators, and the ramped half-and-half method [11] used to generate the initial population, both with maximum individual depth equals to 6. The function set included three binary arithmetic operators ($+$, $-$, $\times$) and the analytic quotient (AQ) [14] as an alternative to the arithmetic division. The terminal set included the variables of the problem and constant values randomly picked from the interval $[-1, 1]$. GSGP employed the geometric semantic crossover for fitness function based on Manhattan distance and mutation operators, as presented in [6], both with probability 0.5.

The base probability (pgd_0) and the decay rate (α) values for all the GD variants are the ones leading to the smaller median training RMSE, as presented in our previous experiments [15]. The values vary in each dataset, as presented in the last two columns of Table 12.1. The two ways of setting the final value of m in the Algorithms 3 and 4, defined by the boolean variable *shiftOne*, were analysed in different configurations. An 'R' in the end of the configuration name indicates that *shiftOne* is `FALSE`, i.e., m is calculated as a random value proportional to the interval nearest to the extreme.

Tables 12.1 and 12.2 present the median training and test RMSE and respective IQR (Interquartile Range), according to 50 executions. Table 12.3 shows the number of datasets were the method in the row is statistically better than the method in the column regarding the test RMSE, according to Wilcoxon test with 95% confidence level. The results indicate the search performed by GSGP benefits from the dispersion provided by the operators, as pointed out by the score of GSGP in relation to the GD configurations. Regarding the use of the shift one algorithm or the random method to compute the values of m in the extremes, there are no significant differences on the dispersion operators. Lastly the results indicate that overall the multiplicative version of the geometric dispersion operator performs better than the additive counterpart.

12.6 Conclusions

This chapter presented a general framework to construct geometric dispersion (GD) operators for GSGP in the context of symbolic regression, followed by two concrete instantiations: the multiplicative geometric dispersion (GD) operator proposed in [15] and another derivation based on the addition operator. These operators move the individuals in order to balance the population around the target output in each dimension of the semantic space, with the objective of expanding the convex hull defined by the population to include the desired output vector.

Experimental analysis was performed on a test bed composed by sixteen datasets to compare the effects of GD operators within GSGP: GSGP with additive

Table 12.2 Test RMSE (median and IQR) obtained by the algorithms for each dataset

Dataset	GSGP		GSGP+A		GSGP+AR		GSGP+M		GSGP+MR	
	Med.	IQR	Med.	IQR	Med.	IQR	Med.	IQR	Med.	IQR
airfoil	8.417	0.757	2.154	0.237	2.131	0.272	2.131	0.243	2.152	0.210
bioavailability	30.736	2.326	31.139	4.170	30.682	4.189	30.860	4.426	30.619	3.773
concrete	5.394	0.642	5.285	0.624	5.244	0.575	5.144	0.635	5.054	0.489
cpu	30.917	15.185	32.804	13.166	32.400	16.182	30.837	14.563	32.027	16.790
energyCooling	1.515	0.147	1.553	0.151	1.489	0.180	1.531	0.159	1.486	0.194
energyHeating	0.956	0.185	0.919	0.157	0.928	0.136	0.971	0.128	0.933	0.169
forestfires	51.632	48.166	52.026	48.626	51.483	50.444	50.227	48.373	50.590	48.762
keijzer-5	0.049	0.005	0.066	0.006	0.065	0.007	0.028	0.009	0.027	0.010
keijzer-6	0.398	0.339	0.275	0.228	0.293	0.203	0.281	0.282	0.250	0.166
keijzer-7	0.018	0.010	0.019	0.010	0.018	0.010	0.017	0.009	0.015	0.008
ppb	28.740	5.290	27.337	5.031	28.139	5.630	28.568	6.170	27.969	5.849
towerData	21.920	1.272	21.979	1.264	21.826	1.263	21.769	1.252	21.871	1.134
vladislavleva-1	0.044	0.030	0.041	0.030	0.046	0.022	0.044	0.030	0.039	0.025
vladislavleva-4	0.052	0.003	0.051	0.003	0.050	0.003	0.051	0.004	0.052	0.002
wineRed	0.620	0.040	0.614	0.041	0.610	0.042	0.615	0.046	0.619	0.049
wineWhite	0.696	0.014	0.695	0.015	0.696	0.014	0.696	0.015	0.696	0.013

Table 12.3 Number of datasets where the method in the row obtained statistically smaller test RMSE in relation to the method in the column

	GSGP	GSGP+A	GSGP+AR	GSGP+M	GSGP+MR	Total (wins)
GSGP	–	1	1	0	0	2
GSGP+A	3	–	1	2	0	6
GSGP+AR	5	1	–	1	4	11
GSGP+M	6	3	5	–	3	17
GSGP+MR	7	4	4	3	–	18
Total (losses)	21	9	11	6	7	

Results according to the Wilcoxon test with 95% confidence level

GD (GSGP+A), multiplicative GD (GSGP+M) and without GD operators were compared regarding the test RMSE. The results showed that GD operators can improve the search performance in terms of test RMSE. Also, they showed that GSGP+M presents advantage over the GSGP+A regarding test RMSE.

Future works include proposing novel dispersion operators following the generic framework, exploring different algorithms to compute the value of m, analysing the impact of using different GD operators simultaneously and tuning the control parameters used by GD operators.

Acknowledgements The authors would like to thank CAPES, FAPEMIG, and CNPq (141985/2015-1) for their financial support.

References

1. Albinati, J., Pappa, G.L., Otero, F.E.B., Oliveira, L.O.V.B.: The effect of distinct geometric semantic crossover operators in regression problems. In: Proceeding of the EuroGP, pp. 3–15 (2015)
2. Banzhaf, W., Nordin, P., Keller, R., Francone, F.: Genetic Programming—An Introduction: On the Automatic Evolution of Computer Programs and Its Applications. Morgan Kaufmann Publishers, San Francisco (1998)
3. Beadle, L., Johnson, C.G.: Semantic analysis of program initialisation in genetic programming. Genet. Program Evolvable Mach. **10**(3), 307–337 (2009)
4. Botzheim, J., Cabrita, C., Kóczy, L.T., Ruano, A.E.: Genetic and bacterial programming for b-spline neural networks design. J. Adv. Comput. Intell. **11**(2), 220–231 (2007)
5. Castelli, M., Vanneschi, L., Silva, S.: Semantic search-based genetic programming and the effect of intron deletion. IEEE Trans. Cybern. **44**(1), 103–113 (2014)
6. Castelli, M., Silva, S., Vanneschi, L.: A C++ framework for geometric semantic genetic programming. Genet. Program Evolvable Mach. **16**(1), 73–81 (2015)
7. Castelli, M., Trujillo, L., Vanneschi, L., Silva, S., Z-Flores, E., Legrand, P.: Geometric semantic genetic programming with local search. In: Proceeding of the GECCO'15, pp. 999–1006. ACM, New York (2015)
8. Castelli, M., Vanneschi, L., Silva, S., Ruberto, S.: How to exploit alignment in the error space: Two different GP models. In: Riolo, R., et al. (eds.) Genetic Programming Theory and Practice XII, Genetic and Evolutionary Computation, pp. 133–148. Springer International Publishing, Cham (2015)

9. Gentle, J.E.: Numerical Linear Algebra for Applications in Statistics. Statistics and Computing. Springer, New York (1998)
10. Gonçalves, I., Silva, S., Fonseca, C.M.F.: On the generalization ability of geometric semantic genetic programming. In: Proceeding of the EuroGP, pp. 41–52 (2015)
11. Koza, J.R.: Genetic Programming: On the Programming of Computers by Means of Natural Selection, vol. 1. MIT Press, Cambridge (1992)
12. Moraglio, A.: Abstract convex evolutionary search. In: Proceeding of the 11th FOGA, pp. 151–162 (2011)
13. Moraglio, A., Krawiec, K., Johnson, C.G.: Geometric semantic genetic programming. In: Proceeding of PPSN XII, vol. 7491, pp. 21–31. Springer, Berlin (2012)
14. Ni, J., Drieberg, R.H., Rockett, P.I.: The use of an analytic quotient operator in genetic programming. IEEE Trans. Evol. Comput. **17**(1), 146–152 (2013)
15. Oliveira, L.O.V.B., Otero, F.E.B., Pappa, G.L.: A dispersion operator for geometric semantic genetic programming. In: Proceeding of the Genetic and Evolutionary Computation Conference, GECCO'16, pp. 773–780. ACM (2016)
16. Pawlak, T.P.: Competent algorithms for geometric semantic genetic programming. Ph.D. thesis, Poznan University of Technology, Poznan (2015)
17. Pawlak, T.P., Krawiec, K.: Semantic geometric initialization. In: Heywood, I.M., McDermott, J., Castelli, M., Costa, E., Sim, K. (eds.) Proceeding of the EuroGP'16. Lecture Notes in Computer Science, vol. 9594, pp. 261–277. Springer International Publishing, Cham (2016)
18. Roman, S.: Advanced Linear Algebra. Graduate Texts in Mathematics, vol. 135, 2nd edn. Springer, New York (2005)
19. Ruberto, S., Vanneschi, L., Castelli, M., Silva, S.: ESAGP - a semantic GP framework based on alignment in the error space. In: Proceeding of the EuroGP, pp. 150–161 (2014)

Chapter 13
Assisting Asset Model Development with Evolutionary Augmentation

Steven Gustafson, Arun Subramaniyan, and Aisha Yousuf

Abstract In this chapter, we explore how Genetic Programming can assist and augment the expert-driven process of developing data-driven models. In our use case, modelers must develop hundreds of models that represent individual properties of parts, components, assets, systems and meta-systems like power plants. Each of these models is developed with an objective in mind, like estimating the useful remaining life or detecting anomalies. As such, the modeler uses their expert judgment, as well as available data to select the most appropriate method. In this initial paper, we examine the most basic example of when the experts select a kind of regression modeling approach and develop models from data. We then use that captured domain knowledge from their processes, as well as end models to determine if Genetic Programming can augment, assist and improve their final results. We show that while Genetic Programming can indeed find improved solutions according to an error metric, it is much harder for Genetic Programming to find models that do not increase complexity. Also, we find that one approach in particular shows promise as a way to incorporate domain knowledge.

Keywords Genetic programming · Lifing models · Machine learning · Industrial applications · Real-world application · Knowledge capture · Artificial intelligence · Intelligent augmentation

---

All authors were employed at GE Global Research, Niskayuna, NY, during the preparation of this chapter.

S. Gustafson (✉)
Maana, Bellevue, WA, USA
e-mail: steven.gustafson@maana.io

A. Subramaniyan
BHGE - Digital, San Ramon, CA, USA

A. Yousuf
Eaton Corporation, Southfield, MI, USA
e-mail: AishaYousuf@eaton.com

R. Riolo et al. (eds.), *Genetic Programming Theory and Practice XIV*, Genetic and Evolutionary Computation,
https://doi.org/10.1007/978-3-319-97088-2_13

13.1 Introduction

One of the initial uses of Genetic Programming was for data modeling. Like other Machine Learning and statistical techniques, data modeling attempts to build a function, or model, that given a set inputs ($\vec{x}$) tries to predict the response variable (y). In the majority of data modeling applications, the model is explanatory, for example with systems identification, as it is built and measured using interpolation. That is, the data from any section, or time, within the data set is used to train and test the model. This is in comparison to extrapolation models, which are built explicitly to predict the future, unseen data points in a world that might change from when the data was collected. In this scenario, insight into the model, what variables it uses, and the soundness of how it is combined to ensure future stability and integration with other system is not just a 'complexity' issue, it is an essential attribute of the modeling process. Genetic Programming and Machine Learning techniques must become assistive to augment the expert's intelligence while building models.

In our primary use case, asset models are built and combined to represent a digital replica of a physical or virtual phenomena for the purpose of providing key insight into optimizations and decisioning. There is a significant effort to validate models and examine their properties from an overall systems integration perspective. There is still a significant challenge to find surprising and valuable data properties within the modeling effort, but the overall error and performance needs to be balanced with the model properties and soundness.

Data modeling with Genetic Programming has been attractive as it easily allows for domain knowledge to be inserted into the search space. For example, it is easy to use different sets of mathematical functions, to determine the appropriate number and ranges of coefficients, or to build in specific routines that are domain unique. In this chapter, we explore how to incorporate domain knowledge in the form of data that has been premanipulated and cleaned by the expert, as well as the expert's final model, which is the result of selecting from several regression techniques and parameters.

13.2 Background

13.2.1 Industrial Asset Models

Models that predict the behavior of complex industrial assets are widely used for both design and fleet management in a variety of industries. For fleet management, the models need to be continually tuned to maintain accuracy and adapted to scale across thousands of assets. In design, asset models are used to compare new designs with previous ones, as well as to explore configuration options when designing for a specific objective, like efficiency or performance, in the light of site or customer specific constraints. Industrial asset models are often created for their capability to forecast and simulate asset behavior in context of other assets and external conditions. The expert modeler, when developing an asset model, is trying to capture the general asset behavior, for example a physics-based understanding of heat transfer, while also capturing the asset specific properties that could be taken advantage of by optimization and trade-off analysis. Given the complexities of both data processing, domain knowledge, and target applications, developing industrial asset models is an expertise intensive task.

13.2.2 *Intelligent Augmentation*

Many Machine Learning and Genetic Programming approaches are used for automating some model building or intelligent task. However, Artificial Intelligence contains a second, less highlighted mission, of augmenting humans with more intelligent systems. In this case, the humans remains "in the loop" rather than being replaced. Augmentative systems are necessary when the problem is so difficult that it can not be fully automated. Augmentative systems are also beneficial when the problem changes and it would be difficult to constantly retrain or tune the underlying Artificial Intelligence system. Thus, the humans and the machines form a partnership where the machines help the humans, and the humans, through their actions of implicit and explicit feedback, help the machines learn to improve. In many cases, data modeling can be considered too difficult to fully automate, and because the problems are constantly changing, it is advantageous to keep the human in the loop so the underlying system can learn and improve. For example, for a class of problems, if the system see the expert consistently reducing the end model to some number of variables, or to reduce the types of operators, the system can begin to search for like models.

Active learning [14] represents the most common human in the loop machine learning systems where a human operator is asked to label or annotate the data during the learning process. Bravo et al. [2] combines machine learning trained on expert, as well as, crowd sourced knowledge with rule-based pattern matching to identify chemical-induced diseases in text. Machine learning is used to classify customer reviews in [17] and supplemented with crowdsourcing when machine learning methods cannot agree on review analysis. The two methods are then combined to come up with the final review classification. Other learning systems in [3] allow intelligent systems to learn via social interaction with humans, as well as self-exploration similar to behavior of children.

In evolutionary searches, prior knowledge can be used to augment the system by starting the search from seeded individuals rather than random initial population. This reduces the search space while improving the convergence time and fitness. Six approaches to seed individuals were tried in [13]: seeding complete prior solutions, seeding parts of prior solutions, randomly rearranged and shuffled versions of prior solutions, and then these three methods with randomized or optimized parameter values from the prior solution. Each one of these methods showed better convergence times and fitness values than randomly generated initial populations. Deep learning methods were used to generate the initial populations in [10] for the genetic programs (GP). The results of the models generated by GP were compared to support vector machines as well as other deep learning methods and showed that seeded GP performed better. In [1] genetic algorithm (GA) is applied to improve the performance of a fuzzy controller. This method uses simulations of the controller designed by a human as an initial population for the GA. In addition, less mutation is applied to the population if the confidence in initial population is high and vice versa. A hybrid approach of seeding individuals in [12] is used for vehicle routing problem to find minimum time to serve all customers. Two clustering methods for clustering the nearest depot and nearest neighbor are used to seed initial population for the GA. Evolutionary algorithms can also be augmented without seeding populations. Statistical measure of attribute quality is used in [11] to augment the GP approach for attribute selection. This insured that good individuals were recombined and reproduced for faster searching of large databases of DNA sequences for predicting diseases. In [9] GP is augmented by considering candidate individuals closely tied to the initial model improved by localized-gradient adaptation. Similarly in [8] the constants in the symbolic expressions are adjusted using a gradient-based nonlinear least squares optimization algorithm.

13.3 Methodology

In this section we describe how asset models are currently developed, specifically for the types developed using data-driven regression modeling approaches. We first describe the hypothesis being tested, and then we motivate the three approaches to leverage a portion of the knowledge embedded in those models coming from both the prepared data as well as the end Regression models. Lastly, we describe the implementation details of those three approaches.

13.3.1 Existing in Model Store

In the modeling framework, subject matter experts from a variety of fields such as thermodynamic performance, material modeling, thermo-mechanical stress analysis, life prediction, reliability, etc., have built thousands of models. These models synthesize decades of deep domain expertise. In many cases, these models contain the unique institutional knowledge that organizations try to leverage for competitive advantage. We have selected regression models built by several experts to predict outputs such as efficiency, damage, stress, etc. to seed this study. The chosen models had 3–30 input variables and ranged from linear to cubic polynomials with several types of features. In addition to being built, a portion of the models were also updated by the experts using Extended or Unscented Kalman filters and Bayesian updating with new sources of data. This provides the evolution of models that can also be used in an automated learning paradigm.

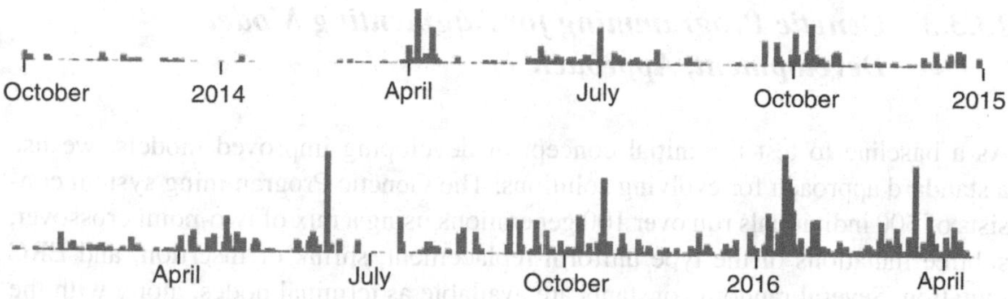

Fig. 13.1 Traffic patterns for industrial asset model building in our specific framework used within a portion of the company. The traffic denotes the frequency of model development tool access by users over several years, demonstrating the significance of expert utilization and the opportunity for expert-augmentation by a technique like Genetic Programming

The models were built over a period of 3 years between 2013 and 2016 with continual engagement from the modelers as shown by the traffic pattern in Fig. 13.1. Significant portion of the modelers were from GE Corporate (53%), GE Aviation (38%), GE Power (5%) and GE Oil & Gas (5%) businesses. The models represent over 50 million rows of data across hundreds of variables. The top modelers have used over 2 million rows to build their models. The average error (root mean squared error) of the models is between 2 and 6% with maximum errors not exceeding 12–15%.

13.3.2 Augmentation Hypothesis

We believe that there is considerable knowledge stored in the built regression models as described above. Knowledge is stored in the dataset that resulted from integrating, filtering and transforming existing data. Knowledge is also stored in the modeling kernel used and its parameters, as well as in the final end result. Many other data, kernel methods and final models could have been created and selected—these final models contained something the expert found meaningful. While many other types of knowledge are leveraged by the domain expert, we hypothesize that Genetic Programming can discover interesting information that would augment the expert during the model development phase. To test this hypothesis, we will first deploy a standard GP system that is primarily leveraging the knowledge stored in the data. To test whether the final selected model is key knowledge, we will define an approach that initializes the entire initial population with variants of the expert's final model. Finally, as a variant of this latter approach, we will initialize the first population using randomly selected subsolutions from the expert's final model.

13.3.3 Genetic Programming for Augmenting Model Development, Approach 1

As a baseline to test the initial concept of developing improved models, we use a standard approach for evolving solutions. The Genetic Programming system consists of 500 individuals run over 100 generations using a mix of two-point crossover, subtree mutations of the type uniform replacement, shrink or insertion, and ERC mutation. Several random constants are available as terminal nodes, along with the standard addition, multiplication, division, subtraction, and power functions. We use a linear scaling approach of the predictions as a way to approximately scale the solution, which is done for each evaluation. A double tournament is used for selection, with a size of 7 individuals, with a likelihood of selecting for fitness 70% of the time. In this approach, individuals are initialized using the standard ramped half-and-half method, with half the solutions generated as full subtrees and half without that constraint, with sizes between 3 and 7. A maximum depth limit of 17 is used. Besides the initialization strategy, all parameters and methods remain the same in the following approaches.

13.3.4 Genetic Programming Mutants, Approach 2

For the following two approaches, we build upon the work of Lipson and Schmidt, specifically we attempt to additionally seed domain knowledge into the process by leveraging the expert-developed model. In the first approach, called Mutants, we simply copy the expert's model 500 times and apply an aggressive mutation strategy. For each solution, we apply with equal probability one of the following: uniformly mutate one subtree to be same size and shape, generate a new subtree using the half-and-half method and replace a random subtree, shrink a random subtree by randomly replacing some of its nodes with its terminals, randomly mutating some of the constants, or do nothing and simply copy the solution. Otherwise, all other parameters and approaches are the same as in the standard approach. The Mutant approach initializes a much larger average solution, dependent upon the size of the expert-developed solution.

13.3.5 Genetic Programming Seeds, Approach 3

In the final approach, which we will call Seeds, we instead look at the expert-developed solution as the source of many potential subtrees to extract and create the initial population. To create the initial population, we randomly select a subtree from the expert-developed solution, and then apply the same mutation strategy as in the Mutant approach, which includes just copying the selected subtree. All other parameters and methods are the same.

13.3.6 Asset Models

We have chosen to use 16 real-world asset models developed to estimate remaining useful life from data (Table 13.1). All the problems were identified as best solved using a regression type model. Of the 16 problems, four problems had 12 variables, four had 3 variables, four had 4 variables, one problem each had 2, 6, 7, and 8 variables. For the number of rows, eight problems used less than 20 variables, seven problems had less than 40 rows, and one had 61 rows. It is important to note that these datasets were created for the purpose of creating a robust model, in this case using a regression technique. In almost all cases, a much, much larger data set existed, but over time as the knowledge of the data and modeling objective increased, the datasets are reduced to the key points from which to build a model. This type of modeling often differs from what traditionally is thought of machine learning, where large amounts of data are used to build "black-box" models. In this case, the process and data curation is more similar to traditional statistical modeling where much more cleaning of the data is done so that the modeler understands very clearly the input data, as well as the model created and its sensitivities.

Although millions of rows of data was available to the modelers, the models identified above use a small subset of highly selective data that the subject matter experts have deemed to be the most valuable. This is unique to industrial systems where large quantities of data is typically available, but is seldom useful because the majority of their operating time is without incident. The events that are unique and need to be modeled are relatively rare. For example, a million flight hours of an aircraft engine might generate a few hundred events that need to tracked and modeled. Tracking every second of operation does not necessarily provide any additional value.

13.3.7 Implementation

The Genetic Programming system was developed using the Distributed Evolutionary Algorithm in Python library [6]. Several custom functions, data ingestion, solution initialization, and output methods were created. In total, about 700 lines of Python code were required to implement the entire system, calling several DEAP and other Python libraries.

To capture the asset models, we leveraged a novel, cloud-based analytic platform that enables model development, as well as captures meta data about the models developed. The meta data was modeled and normalized across expert activity using an open source, domain specific language [7] we developed to write and manage

Table 13.1 The sixteen asset models, all coefficients are masked as letter C, and variable names are represented by a unique name starting with V

Number	Model
1	$C \times C + ((V1 - C) \div C) \wedge C \times C + ((V2 - C) \div C) \wedge C \times C$
2	$C \times C + ((V1 - C) \div C) \wedge C \times -C + ((V2 - C) \div C) \wedge C \times C + ((V3 - C) \div C) \wedge C \times -C + ((V1 - C) \div C) \wedge C \times C + ((V1 - C) \div C) \wedge C \times ((V3 - C) \div C) \wedge C \times -C + ((V3 - C) \div C) \wedge C \times -C + ((V1 - C) \div C) \wedge C \times ((V3 - C) \div C) \wedge C \times C + ((V1 - C) \div C) \wedge C \times ((V3 - C) \div C) \wedge C \times C$
3	$C \times C + ((V1 - C) \div C) \wedge C \times C + ((C - C) \div C) \wedge C \times ((V2 - C) \div C) \wedge C \times C$
4	$C \times C + ((V1 - C) \div C) \wedge C \times C$
5	$C \times C + ((V1 - C) \div C) \wedge C \times C + ((V1 - C) \div C) \wedge C \times ((V2 - C) \div C) \wedge C \times C$
6	$C \times C + ((V1 - C) \div C) \wedge C \times C + ((V2 - C) \div C) \wedge C \times C + ((V3 - C) \div C) \wedge C \times C + ((V2 - C) \div C) \wedge C \times C + ((V3 - C) \div C) \wedge C \times -C$
7	$C \times C + ((V1 - C) \div C) \wedge C \times C + ((V1 - C) \div C) \wedge C \times -C + ((V2 - C) \div C) \wedge C \times ((V3 - C) \div C) \wedge C \times C + ((V2 - C) \div C) \wedge C \times ((CA - C) \div C) \wedge C \times C + ((V3 - C) \div C) \wedge C \times ((V1 - C) \div C) \wedge C \times -C$
8	$C \times C + ((V1 - C) \div C) \wedge C \times C + ((V1 - C) \div C) \wedge C \times -C + ((V1 - C) \div C) \wedge C \times C$
9	$C \times C + ((V1 - C)/C) \wedge C \times -C$
10	$C \times C + ((V1 - C) \div C) \wedge C \times C + ((V2 - C) \div C) \wedge C \times -C + ((V5 - C) \div C) \wedge C \times C + ((V3 - C) \div C) \wedge C \times -C + ((V1 - C) \div C) \wedge C \times ((V6 - C) \div C) \wedge C \times -C + ((V1 - C) \div C) \wedge C \times ((V2 - C) \div C) \wedge C \times -C + ((V1 - C) \div C) \wedge C \times ((V5 - C) \div C) \wedge C \times C + ((V6 - C) \div C) \wedge C \times ((V4 - C) \div C) \wedge C \times C + ((V6 - C) \div C) \wedge C \times ((V3 - C) \div C) \wedge C \times C + ((V6 - C) \div C) \wedge C \times C$
11	$((V1 - C) \div C) \wedge C \times ((V2 - C) \div C) \wedge C \times C + ((V1 - C) \div C) \wedge C \times ((V3 - C) \div C) \wedge C \times -C + ((V1 - C) \div C) \wedge C \times ((V4 - C) \div C) \wedge C \times -C + ((V3 - C) \div C) \wedge C \times ((V4 - C) \div C) \wedge C \times -C$
12	$C \times C + ((V1 - C) \div C) \wedge C \times C + ((V2 - C) \div C) \wedge C \times C + ((V3 - C) \div C) \wedge C \times -C + ((V2 - C) \div C) \wedge C \times ((V3 - C) \div C) \wedge C \times C + ((V3 - C) \div C) \wedge C \times -C + ((V2 - C) \div C) \wedge C \times ((V3 - C) \div C) \wedge C \times C$
13	$C \times -C + ((V1 - C) \div C) \wedge C \times -C + ((V2 - C) \div C) \wedge C \times -C + ((V3 - C) \div C) \wedge C \times C + ((V4 - C) \div C) \wedge C \times C + ((V5 - C) \div C) \wedge C \times -C$
14	$C \times C + ((V1 - C) \div C) \wedge C \times -C + ((V2 - C) \div C) \wedge C \times C + ((V3 - C) \div C) \wedge C \times -C + ((V4 - C) \div C) \wedge C \times -C + ((V2 - C) \div C) \wedge C \times ((V3 - C) \div C) \wedge C \times -C + ((V3 - C) \div C) \wedge C \times ((V4 - C) \div C) \wedge C \times C$
15	$C \times C + ((V1 - C) \div C) \wedge C \times C$
16	$C \times C + ((V1 - C) \div C) \wedge C \times C + ((V2 - C) \div C) \wedge C \times -C + ((V3 - C) \div C) \wedge C \times C + ((V4 - C) \div C) \wedge C \times C + ((V2 - C) \div C) \wedge C \times C + ((V4 - C) \div C) \wedge C \times -C$

ontologies [5] that leverages the OWL format [15]. A novel Big Data technology to ingest, integrate, and query data in a visual and Artificial Intelligence assisted way [16]. In this way, we are able to monitor the experts but also have a platform on which to provide automated recommendations and augmentative intelligent systems, where semantic technologies provide a key approach for experts to manage and understand data [4].

To create the empirical results, we leveraged a big data cluster consisting of over 1000 cores and 10 TB of RAM. Across the 16 problems, we performed 30 random runs, completing in less than an hour using a Map Reduce implementation. While

Table 13.2 The ability of the Genetic Programming approaches (Standard GP, Mutant GP and Seed GP) to evolve a model comparable or better to the expert model is measured below when the RMSE and Solution complexity is better (E&S≤), when the RMSE is better (E≤), and when the solution complexity is better (S≤)

	Standard GP				Mutant GP				Seed GP			
Model	E&S≤	S≤	E≤	Ave Gen	E&S≤	S≤	E≤	Ave Gen	E&S≤	S≤	E≤	Ave Gen
1	0	1	8	93.0	0	0	30	99.0	0	0	27	96.9
2	3	15	9	100.4	11	11	30	98.3	1	24	2	98.4
3	0	0	30	99.5	0	0	30	98.4	20	20	30	65.4
4	0	0	30	99.4	0	0	30	99.0	0	0	30	100.2
5	1	1	30	97.0	0	0	30	97.7	20	20	30	61.0
6	0	4	21	94.4	17	17	30	74.5	1	5	16	98.3
7	1	1	29	100.2	3	3	30	98.6	7	17	18	80.9
8	1	1	30	99.4	2	2	30	95.7	2	2	30	96.9
9	0	0	30	100.0	0	0	30	95.4	0	3	24	89.3
10	11	29	11	97.1	18	23	25	97.1	6	28	6	99.7
11	6	6	30	92.9	11	11	30	97.5	13	13	30	95.1
12	0	24	0	78.3	0	16	0	97.0	0	30	0	54.4
13	0	7	9	96.3	0	3	1	95.9	0	13	1	81.5
14	0	21	0	77.3	0	18	0	94.0	0	30	0	46.2
15	0	1	16	97.4	0	0	30	98.5	0	14	10	83.1
16	0	5	0	99.2	0	6	0	99.1	0	25	0	73.2
Totals	23	116	283	**95.11**	62	110	356	**95.97**	70	244	254	**82.54**

The average generation when the best solution is found is reported in (Ave Gen); the bold-faced values in the last row (Total) of this column are *averages of the averages* in the column above

possible, we did not implement the DEAP multiprocessing capability, which could have further improved the results.

13.4 Results

From Table 13.2 it can be seen that Genetic Programming is able to find models that improve over the expert built models in many cases. Even in cases where the model accuracy did not improve significantly, the model forms changed substantially. Table 13.3 shows one selected best model from the many, many candidates found during the evolutionary search. In many cases, the new model forms identified by the GP process help provide new insights to the modeler and can be used to accelerate the model building process.

A specific example of a user generated model that was used to seed the GP process is shown below.

$$y = f(X_1 + X_2) + g(X_1 \times X_2) \tag{13.1}$$

One of the best models selected by GP by using the above model is shown below.

$$y = \eta\left(\frac{X_2}{X_1}\right) + \psi\left(X_1 + \phi(X_2)\right) \qquad (13.2)$$

13.4.1 Performance Alone

To measure performance, we computed the expert model's normalized root mean squared error (RMSE) and compared it to the Genetic Programming produced model. If the Genetic Programming model was within 10% of the expert model's normalized RMSE, we considered that a success.

Standard Genetic Programming was able to produce a successful error as compared to the expert for 13 of the 16 problems. The same result, on the same problems, was found for the Mutant and Seed approaches. Three of the problems proved too difficult for Genetic Programming to find an improved error. For standard Genetic Programming, on average 22.8 of the 30 runs resulted in an improved error, but Mutants had successful errors on 29.5 of the 30 runs. The Seed method found a success on an average of 21.1 runs out of 30. Thus, the Mutant method was slightly more successful in finding a successful error.

13.4.2 Solution Complexity Alone

To measure solution complexity, we counted the number of nodes (operators, coefficients and variables) in a solution. If the Genetic Programming produced model was within 20% of the expert model, we considered that a success.

For standard Genetic Programming, it was able to find successful solutions relative to size for 13 problems, with Mutants and Seed being successful for 10 and 14 problems respectively. Only on one problem did none of the approaches find a successful solution with respect to size. However, the Seed approach had a success solution on average 17.4 runs out of 30, with Mutants at 11 runs and standard Genetic Programming with 8.9 runs being successful on average. While the distributions would tell more information about this success, the average does indicate that the Seed approach was more capable of finding less complex solutions.

13.4.3 Performance and Complexity

Lastly, for a Genetic Programming solution to be viable, we required that it achieved success in both the normalized RMSE and size requirements as above.

For standard Genetic Programming, only 6 of the 16 problems were successfully solved, and the same was true of Mutants. Interestingly, only 5 of the 6 were the same between the two approaches. The Seed approach solved those same 6 plus 2 additional problems successfully. For those problems where a solution was found, standard Genetic Programming had on average 3.8 runs out of 30 successful, Mutants and Seed both found success on average of 10 runs out of 30. The Seed approach was able to leverage the expert model to find both good error, as well as low complexity solutions.

While the results and comparisons are all statistically sound, the fact that in this study only 16, albeit real-world, problems were studied, an no in-depth optimization of the underlying Genetic Programming approaches were performed, one should take caution in extrapolating too much toward general behavior at this time. We can conclude that we have found a viable way to augment the expert-driven modeling process by finding viable, alternative models to standard regression models by automatically using the expert domain knowledge embedded in their data and their developed models.

13.4.4 Unique Insights from GP Process

Consider the equations derived from GP based on user generated seed models. It is interesting to note that the expert modeler chose only the product of the two variables to be important. However, the GP process correctly identified that the ratio of the variables is more important than the product. This was verified by the subject matter experts as well. This kind of information is very valuable in reducing model building time and potentially automating the entire process to derive broad business benefits.

On the other end of the spectrum, there were cases where the modeler tried building sophisticated models with several terms and the GP process correctly identified that there was no sufficient data to justify the model complexity. This prevents overtuning or mistuning models where data is sparse.

Thus, both feature identification and model simplification is possible in addition to model performance improvements with GP.

13.4.5 Future Enhancements

There are several next steps and future enhancements we are pursuing.

Firstly, the methods we are comparing Genetic Programming to all perform some form of coefficient optimization. This will be a relatively simple capability to add into the Genetic Programming approach, albeit a potentially computational-costly one. Therefore, we plan on optimizing the Genetic Programming algorithm and

Table 13.3 For each of the sixteen asset models, we show below one of the many better models found by the Seed Genetic Programming approach. There were many potential candidate models found by the other Genetic Programming approaches, but for illustrative purposes, we selected one of the best from the final generation. All coefficients are masked as letter C, and variable names are represented by a unique name starting with V. Note that the masked variables do not correspond to the original asset models in Table 13.1, and that substitution ignored unary negative signs which have not been removed by arithmetic simplification

Number	Model
1	$((((((V1 - (V1 \div ((((((V1 - (((C \times C) + V1) \div C)) \div V2) + (V1 - C)) - C) - C) + V1))) \div V2) + (V1 - C)) - C) \div C) \wedge C \times ((((-V1 - V1) - V1) \div V2) - C))$
2	$-(((V1 \times -(V1 \times V2)) - ((((V1 \times (V1 \times V2)) - -V2) - V1) \div (((V2 + (V2 \div -V3)) + (V2 \div -V3)) + (V2 \div -V3)))) \div ((V2 + (V2 \div -V3)) - ((-(V2 - ((V2 - (-V2 - V1)) \div ((V1 \times -(V1 \times (V1 \times ((V1 \times V2) + V2)))) - (-V2 - (-V2 \div V2)))))) - (-V2 - V1)) \div V2)))$
3	$(((C + V1) \times C) \div ((V2 + ((((V1 \div (V2 + C)) + V1) \div C) \div (V2 \div C))) \div C))$
4	$(C + -(((V1 - (C + ((((V1 - ((C + (V2 \div C)) + (C \div C))) \div C) \times ((C + (C + (C \div C))) + C)) \div (V2 \div C)))) \div C) \wedge C \times (C + C)))$
5	$(((V1 - V2) - V3) \div -(V1 - V2))$
6	$(C + ((((V1 - C) \div C) \times C) + ((((V2 - C) \div C) \times C) + ((((V3 - C) \div C) \times C) + ((((V2 - C) \div C) \wedge C \times C) + (((V3 - C) \div C) \wedge C \times -C))))))$
7	$((((V1 - C) \div V1) \div ((((V1 - V1 \wedge (V1 \times V1)) - ((V1 - C) \div V1)) \div C) \times V1)) - (((C \div C) - C) \div C))$
8	$((C \div V1) - (((V2 - V5) - (((V5 - ((V2 \div C) + (V4 + V2))) + ((V6 - (C \div V1)) + V3)) \div V1) \wedge C) + V3))$
9	$(((((V1 - C) \div C) \wedge C \times C) + (C \times -(-((((((V1 - C) \div C) \wedge C \div -(((V1 - C) - C) \div (V3 \div C))) \wedge C \times V1) \div V2) \wedge C \times -C) \wedge C \times -C))) \times V2)$
10	$-(((V1 + V3) \div (V4 + ((V5 \times (((V1 + V2) + -V2) - V5)) + (((V1 + ((((V5 \div V6) - ((V5 + V5) + (V5 \times V5))) + ((- - (V4 + V1) \wedge - - V2 + (V4 + V4)) \div (- - V2 + (-V5 - V6)))) \times V6)) + V5) \div V1)))) - (((V4 + V1) \div (V2 + -V1)) \div V1))$
11	$(C + (((V1 - ((((V2 - C) \div C) \wedge C) \wedge C \times C)) \div C) \times ((V2 - ((((V1 - (C - V3 \wedge C)) - ((C - V3 \wedge C) - V3 \wedge C)) - (((V2 - C) \div C) \wedge (C - C) \wedge (C \times (C - C)) \times C)) \div C)) \div C)))$
12	$((V1 + (V1 + (((((V1 + V1) - (V1 \div ((V2 + V4) - V1))) - (((V3 + V4) - V1) \div V4)) - (V4 \div V1)) + V1))) \div V1)$
13	$(((-((V2 - (V1 \div V6)) + -((V7 - V8) - V3)) \times (V3 \div V8)) + (-(V4 - (V8 - ((V7 - -(((V2 + V2) \div (V7 - V3)) - (V8 - ((((V2 \div ((V7 - V4) - V4)) + (V8 + (V5 + V4))) + ((V6 - V7) - V8)) \div (V3 - V9))))) - V4))) + (V2 + -((V7 - V8) - V4)))) \div -V5)$
14	$(C + ((-(C \div C) \div (((C \div C) \wedge C \times (C + C)) + (((((((C + C) \div (C - (V1 - C))) \wedge C \times C) + ((-(C \div C) \div ((-C \div V2) - (V1 - C))) \wedge C + -C \wedge C)) - C) \div C) \div ((V1 - C) \div V1)) + (((-C \div V2) \wedge C \times C) + (((-C \div V2) \wedge C \times C) + -C))))) + -(C \times C)))$
15	$-((C - (C - -(-(V1 \div C) \div ((C + C) \times (C \div C))))) - (C - (C - C)))$
16	$((((V1 - ((V7 \div -((V1 + (((V4 + -V6) + V4) \div (V5 - V6))) + ((-V1 - ((V7 \div -((V1 + (-V2 \div V1)) + ((-V6 \div V7) - - - V7))) \times V4)) \div (V4 + -V7)))) \times V4)) + (V3 \div ((V7 \div -(-((V1 + (V4 \div (V4 + V4))) + (V1 - - - (V4 \div V4))) + (((V1 + (V4 \div V4)) \times (V2 + V2)) \div (V4 + -(V1 + V5))))) \ast V7))) \div (V4 + V4)) \div V2)$

testing for best parameters to see how small the population can be, as well as identify a good stopping criterion.

Secondly, we plan on building an augmentation module into our model development process so as to provide guidance to the users using the Genetic Programming produce models. In this work, we will explore the best user experiences as well as the feedback from whether the users found the Genetic Programming solutions helpful or not. Several possible integrations are possible. For example, we could present the users with full, complete solutions. Or, we could identify common subtrees from the best solutions and present them as potential new features for the users to compute, even going so far as to prepare new data sets with those features represented.

13.5 Conclusions

In this chapter, we presented the use case of the developing an augmentative approach to developing industrial asset models. Using 16 real-world asset models, we explored how domain knowledge can be exploited by Genetic Programming. We demonstrated that using the domain knowledge from the data alone enables Genetic Programming to find potentially interesting solutions. This standard result is novel here as the data used has been highly reduced to fewer number of rows and columns to enable robust regression models to be developed by the expert. Genetic Programming in this manner, as typical, tends to produce much larger solutions than the expert developed model. When Genetic Programming is able to leverage the domain knowledge in the data as well as from the expert developed model, in the two approaches Mutants and Seeds described here, it is capable of finding solutions with good or similar performance as well as smaller or similar complexity. The Seeds Genetic Programming approach, which initiates the population with randomly extracted subtrees that are then mutated, was the most effective in finding lower complexity solutions that also had good error performance.

References

1. Akbarzadeh-T., M.R., Jamshidi, M.: Incorporating a-priori expert knowledge in genetic algorithms. In: 1997 IEEE International Symposium on Computational Intelligence in Robotics and Automation, 1997. CIRA'97. Proceedings, pp. 300–305 (1997)
2. Bravo, A., Li, T., Su, A.I., Good, B.M., Furlong, L.: Combining machine learning, crowdsourcing and expert knowledge to detect chemical-induced diseases in text. In: Proceedings of the Fifth BioCreative Challenge Evaluation Workshop, pp. 266–273 (2015)
3. Breazeal, C., Thomaz, A.L.: Learning from human teachers with socially guided exploration. In: IEEE International Conference on Robotics and Automation, 2008. ICRA 2008, pp. 3539–3544 (2008)

4. Crapo, A., Gustafson, S.: Semantics: revolutionary breakthrough or just another way of doing things? In: Semantic Web: Implications for Technologies and Business Practices, pp. 85–118. Springer International Publishing, Cham (2016)
5. Crapo, A., Moitra, A.: Toward a unified english-like representation of semantic models, data, and graph patterns for subject matter experts. Int. J. Semantic Comput. **07**(03), 215–236 (2013). https://doi.org/10.1142/S1793351X13500025
6. Fortin, F.A., De Rainville, F.M., Gardner, M.A., Parizeau, M., Gagné, C.: DEAP: evolutionary algorithms made easy. J. Mach. Learn. Res. **13**, 2171–2175 (2012)
7. GE Global Research: Semantic Application Design Language (SADL): Open Source Project on Source Forge (2011). http://sadl.sourceforge.net/sadl.html
8. Kommenda, M., Kronberger, G., Winkler, S., Affenzeller, M., Wagner, S.: Effects of constant optimization by nonlinear least squares minimization in symbolic regression. In: Proceedings of the 15th Annual Conference Companion on Genetic and Evolutionary Computation, GECCO '13 Companion, pp. 1121–1128. ACM, New York (2013)
9. La Cava, W.G., Danai, K.: Gradient-based adaptation of continuous dynamic model structures. Int. J. Syst. Sci. **47**(1), 249–263 (2016)
10. Lu, Q., Ren, J., Wang, Z.: Using genetic programming with prior formula knowledge to solve symbolic regression problem. Comput. Intell. Neurosci. **2016**, 17 (2016)
11. Moore, J.H., White, B.C.: Exploiting expert knowledge in genetic programming for genome-wide genetic analysis. In: Runarsson, T.P., et al. (eds.) Parallel Problem Solving from Nature - PPSN IX: 9th International Conference, Reykjavik, September 9–13, 2006, Proceedings, pp. 969–977. Springer, Berlin (2006)
12. Sathyanarayanan, S., Joseph, K.S., Jayakumar, S.K.V.: A hybrid population seeding technique based genetic algorithm for stochastic multiple depot vehicle routing problem. In: 2015 International Conference on Computing and Communications Technologies (ICCCT), pp. 119–127 (2015)
13. Schmidt, M.D., Lipson, H.: Incorporating expert knowledge in evolutionary search: a study of seeding methods. In: Proceedings of the 11th Annual Conference on Genetic and Evolutionary Computation, GECCO '09, pp. 1091–1098. ACM, New York (2009)
14. Settles, B.: Active learning literature survey. Tech. Rep. Report 1648, University of Wisconsin, Madison (2010)
15. W3C OWL Working Group: OWL Web Ontology Language Reference. W3C Recommendation (2004). http://www.w3.org/TR/owl-ref/
16. Williams, J.W., Cuddihy, P., McHugh, J., Aggour, K.S., Menon, A., Gustafson, S., Healy, T.: Semantics for big data access integration: improving industrial equipment design through increased data usability. In: 2015 IEEE International Conference on Big Data (Big Data), pp. 1103–1112 (2015)
17. Wu, H., Sun, H., Fang, Y., Hu, K., Xie, Y., Song, Y., Liu, X.: Combining machine learning and crowdsourcing for better understanding commodity reviews. In: Proceedings of the Twenty-Ninth AAAI Conference on Artificial Intelligence, AAAI'15, pp. 4220–4221. AAAI Press, San Francisco (2015)

Chapter 14
Identifying and Harnessing the Building Blocks of Machine Learning Pipelines for Sensible Initialization of a Data Science Automation Tool

Randal S. Olson and Jason H. Moore

Abstract As data science continues to grow in popularity, there will be an increasing need to make data science tools more scalable, flexible, and accessible. In particular, automated machine learning (AutoML) systems seek to automate the process of designing and optimizing machine learning pipelines. In this chapter, we present a genetic programming-based AutoML system called TPOT that optimizes a series of feature preprocessors and machine learning models with the goal of maximizing classification accuracy on a supervised classification problem. Further, we analyze a large database of pipelines that were previously used to solve various supervised classification problems and identify 100 short series of machine learning operations that appear the most frequently, which we call the *building blocks* of machine learning pipelines. We harness these building blocks to initialize TPOT with promising solutions, and find that this sensible initialization method significantly improves TPOT's performance on one benchmark at no cost of significantly degrading performance on the others. Thus, sensible initialization with machine learning pipeline building blocks shows promise for GP-based AutoML systems, and should be further refined in future work.

Keywords Pipeline optimization · Hyperparameter optimization · Automated machine learning · Sensible initialization · Building blocks · Genetic programming · Pareto optimization · Multiobjective optimization · Data science · Python language

R. S. Olson (✉) · J. H. Moore
Institute for Biomedical Informatics, University of Pennsylvania, Philadelphia, PA, USA
e-mail: rso@randalolson.com; jhmoore@upenn.edu

R. Riolo et al. (eds.), *Genetic Programming Theory and Practice XIV*, Genetic and Evolutionary Computation,
https://doi.org/10.1007/978-3-319-97088-2_14

14.1 Introduction

Machine learning is often touted as a "field of study that gives computers the ability to learn without being explicitly programmed" [26]. Despite this common claim, it is well-known by machine learning practitioners that designing effective machine learning pipelines is often a tedious endeavor, and typically requires considerable experience with machine learning algorithms, expert knowledge of the problem domain, and brute force search to accomplish. Figure 14.1 depicts a typical machine learning pipeline, where each step requires intervention by machine learning practitioners. Thus, contrary to what machine learning enthusiasts would have us believe, machine learning still requires considerable explicit programming.

In response to this challenge, several automated machine learning methods have been developed over the years [14]. Over the past year, we have been developing a Tree-based Pipeline Optimization Tool (TPOT) that automatically designs and optimizes machine learning pipelines for a given problem domain [21], without any need for human intervention. In short, TPOT optimizes machine learning pipelines using a version of genetic programming (GP), a well-known evolutionary computation technique for automatically constructing computer programs [1, 16]. Previously, we demonstrated that combining GP with Pareto optimization enables TPOT to automatically construct high-accuracy *and* compact pipelines that consistently outperform basic machine learning analyses [20]. In this chapter, we report on our progress toward introducing sensible initialization [12] of the GP population into TPOT, with the goal of enabling TPOT to harness expert knowledge about machine learning pipelines to efficiently discover effective pipelines for a given problem domain.

Previous research on the initialization of GP populations has shown that the initialization process can vitally affect the performance of GP algorithms [9, 18, 22].

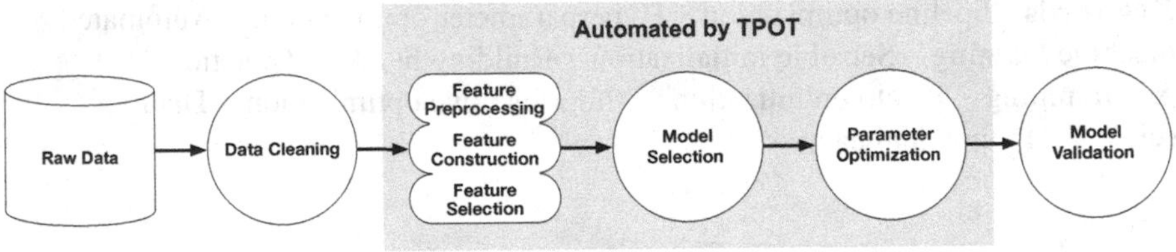

Fig. 14.1 A typical machine learning pipeline. Machine learning practitioners often start with a raw data set that must be formatted, have missing values imputed, and otherwise prepared for analysis. Following this step, practitioners must often transform the feature set into a format that is amenable to modeling, for example by preprocessing the features via scaling, constructing new features from existing features, or removing less useful features via feature selection. Next, practitioners must select a machine learning model to fit to the data, then optimize the parameters of the model and feature transformation operations to allow the model to best capture the underlying signal in the data. At the end of this process, practitioners must evaluate their pipeline on a validation data set that the pipeline never saw before, which allows the practitioners to determine whether the pipeline generalizes beyond the initial training data

However, most of this research has focused on generating a diversity of valid GP tree structures, which may not be useful in all application domains. Here we follow in the footsteps of [12] and focus on harnessing expert knowledge—in this case, expert knowledge about machine learning pipelines—to initialize the GP population. In particular, we attempt to identify the building blocks [10] of machine learning pipelines, and harness these building blocks for sensible initialization of the GP population in TPOT.

14.2 Automated Machine Learning

In the early days of machine learning automation research (AutoML for short), researchers focused primarily on hyperparameter optimization [14]. For example, the most commonly-used form of hyperparameter optimization is *grid search*, where users apply brute force search to evaluate a predefined range of model parameters to find the model parameters that allows for the best model fit. More recently, researchers showed that it is possible to discover optimal parameter sets faster than exhaustive grid search by randomly sampling within a predefined grid search [2], which shows promise for guided search in the hyperparameter space. Bayesian optimization, in particular, has proven effective for hyperparameter optimization and has even outperformed manual hyperparameter tuning by expert practitioners [27].

Another focus of AutoML research has been feature construction. One recent example of automated feature construction is the "Data Science Machine," which automatically constructs features from relational databases via deep feature synthesis. In their work, [15] demonstrated the crucial role of automated feature construction in machine learning pipelines by entering their Data Science Machine in three machine learning competitions and achieving expert-level performance in all of them. Thus, we know that automated feature construction can play a vital role in AutoML systems.

More recently, Feurer et al. [6] developed an AutoML system called *auto-sklearn*, which uses Bayesian optimization to discover the ideal combination of data and feature preprocessors, models, and model hyperparameters to maximize classification accuracy for a particular problem domain. However, auto-sklearn optimizes over a predefined set of pipelines that only include one data preprocessor, one feature preprocessor, and one model, which precludes auto-sklearn from producing arbitrarily large pipelines that may be important for AutoML.

Zutty and colleagues [30] demonstrated an AutoML system using genetic programming (GP) to optimize machine learning pipelines for signal processing, and found that GP is capable of designing better pipelines than humans for one signal processing task. As such, GP shows considerable promise in the AutoML domain, and we significantly extend this work with TPOT.

14.3 Methods

In the following sections, we provide an overview of the Tree-based Pipeline Optimization Tool (TPOT), including the machine learning operators used as genetic programming (GP) primitives, the tree-based pipelines used to combine the primitives into working machine learning pipelines, and the GP algorithm used to evolve said tree-based pipelines. We further describe the process we implemented to provide sensible initialization for the GP algorithm in TPOT, and conclude with a description of the data sets used to evaluate this new version of TPOT. TPOT is an open source project on GitHub, and the underlying Python code can be found at https://github.com/rhiever/tpot.

14.3.1 Machine Learning Pipeline Operators

At its core, TPOT is a wrapper for the Python machine learning package, scikit-learn [23]. Thus, each machine learning pipeline operator (i.e., GP primitive) in TPOT corresponds to a machine learning algorithm, such as a supervised classification model. All implementations of the machine learning algorithms listed below are from scikit-learn (except XGBoost), and we refer to the scikit-learn documentation [23] and [13] for detailed explanations of the machine learning algorithms used in TPOT.

Supervised Classification Operators DecisionTree, RandomForest, eXtreme Gradient Boosting Classifier (from XGBoost, [4]), LogisticRegression, and KNearestNeighborClassifier. Classification operators store the classifier's predictions as a new feature as well as the classification for the pipeline.

Feature Preprocessing Operators StandardScaler, RobustScaler, MinMaxScaler, MaxAbsScaler, RandomizedPCA [19], Binarizer, and PolynomialFeatures. Preprocessing operators modify the data set in some way and return the modified data set.

Feature Selection Operators VarianceThreshold, SelectKBest, SelectPercentile, SelectFwe, and Recursive Feature Elimination (RFE). Feature selection operators reduce the number of features in the data set using some criteria and return the modified data set.

We also include an operator that combines disparate data sets, as demonstrated in Fig. 14.2, which allows multiple modified copies of the data set to be combined into a single data set. Lastly, we provide integer and float terminals to parameterize the various operators, such as the number of neighbors (k) in the k-Nearest Neighbors Classifier.

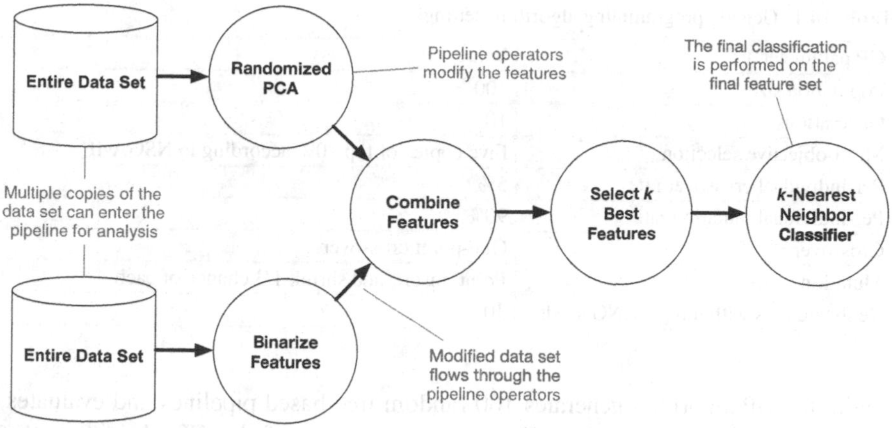

Fig. 14.2 An example tree-based pipeline from TPOT. Each circle corresponds to a machine learning operator, and the arrows indicate the direction of the data flow

14.3.2 Constructing Tree-Based Pipelines

To combine these operators into a machine learning pipeline, we treat them as GP primitives and construct GP trees from them. Figure 14.2 shows an example tree-based pipeline, where two copies of the data set are provided to the pipeline, modified in a successive manner by each operator, combined into a single data set, and finally used to make classifications. Because all operators receive a data set as input and return the modified data set as output, it is possible to construct arbitrarily large machine learning pipelines that can act on multiple copies of the data set. Thus, GP trees provide an inherently flexible representation of machine learning pipelines.

In order for these tree-based pipelines to operate, we store three additional variables for each record in the data set. The "class" variable indicates the true label for each record, and is used when evaluating the accuracy of each pipeline. The "guess" variable indicates the pipeline's latest guess for each record, where the classifications from the last classification operator in the pipeline are stored as the "guess". Finally, the "group" variable indicates whether the record is to be used as a part of the internal training or testing set, such that the tree-based pipelines are only trained on the training data and evaluated on the testing data. We note that the data set provided to TPOT is split into an internal stratified 75%/25% training/testing set.

14.3.3 Optimizing Tree-Based Pipelines

To automatically generate and optimize these tree-based pipelines, we use a GP algorithm [1, 16] as implemented in the Python package DEAP [8]. The TPOT GP algorithm follows a standard GP process with settings listed in Table 14.1. To

Table 14.1 Genetic programming algorithm settings

GP parameter	Value
Population size	100
Generations	100
Multi-objective selection	Five copies of top 20% according to NSGA-II
Per-individual crossover rate	5%
Per-individual mutation rate	90%
Crossover	One-point crossover
Mutation	Point, insert, and shrink 1/3 chance of each
Replicate runs with unique RNG seeds	30

begin, the GP algorithm generates 100 random tree-based pipelines and evaluates their accuracy on the data set. For every generation of the GP algorithm, the algorithm selects the top 20 pipelines in the population according to the NSGA-II selection scheme [5], where pipelines are selected to simultaneously maximize classification accuracy on the data set while minimizing the number of operators in the pipeline. Each of the top 20 selected pipelines produce five offspring into the next generation's population, 5% of those offspring experience crossover with another offspring, then 90% of the remaining unaffected offspring experience random mutations. Every generation, the algorithm updates a Pareto front of the non-dominated solutions [5] discovered at any point in the GP run. The algorithm repeats this evaluate-select-crossover-mutate process for 100 generations—adding and tuning pipeline operators that improve classification accuracy and pruning operators that degrade classification accuracy—at which point the algorithm selects the highest-accuracy pipeline from the Pareto front as the representative "best" pipeline from the run.

14.3.4 Sensible Initialization in TPOT

Next, we implement a version of TPOT with sensible initialization, which we call *TPOT-SI*. In TPOT-SI, the GP algorithm creates the initial population by seeding it with a random selection from 100 building blocks that we identified from previous TPOT runs. These building blocks consist of tree-based pipelines with 1–3 operators, e.g., one building block that we identified was "PolynomialFeatures → LogisticRegression," where the building block casts the data set into a polynomial feature space then provides those features to a logistic regression model to make the classification.

The primary idea behind providing sensible initialization via building blocks is that genetic programming algorithms typically rely heavily on crossover [24]. Thus, we want to initialize the GP population with building blocks that (1) start the GP population off with pipelines that already effectively solve at least part of the classification task, and (2) can be mixed and matched to build better pipelines in a more efficient manner. However, we note that our preliminary investigations showed

Table 14.2 Ten most frequent machine learning pipeline building blocks

RandomForest
XGBClassifier
LogisticRegression
DecisionTree
KNearestNeighborClassifier
XGBClassifier → RandomForest
LogisticRegression → RandomForest
PolynomialFeatures → LogisticRegression
PolynomialFeatures → RandomForest
SelectPercentile → RandomForest

that a high crossover regime (per-individual crossover rate = 90%, per-individual mutation rate = 5%) performed worse than a high mutation regime (per-individual crossover rate = 5%, per-individual mutation rate = 90%) for both TPOT and TPOT-SI, so we focused on the high mutation regime in this chapter.

To identify these building blocks, we ran 30 replicates of TPOT on the 160 supervised classification benchmark data sets described in Sect. 14.3.5. We identified the highest-accuracy tree-based pipeline from the final Pareto front for each run, then performed an "n-gram" analysis (up to $n = 4$) of the pipelines to count the most frequent combinations of pipeline operators. For example, in Fig. 14.2 "SelectKBest → KNearestNeighborClassifier" would be a 2-gram, and "KNearestNeighborClassifier" would be a 1-gram. After counting all of the n-grams in each tree-based pipeline, we summed the counts across all 4800 replicates to determine the 100 most frequent n-grams, which we used as TPOT building blocks.[1] We have listed the top ten most frequent building blocks in Table 14.2.

14.3.5 Benchmark Data

We compiled 160 supervised classification benchmarks[2] from a wide variety of sources, including the UCI machine learning repository [17], a large preexisting benchmark repository from [25], and simulated genetic analysis data sets from [28]. These benchmark data sets range from 60 to 60,000 records, few to hundreds of features, and include binary as well as multi-class supervised classification problems. We selected data sets from a wide range of application domains, including genetic analysis, image classification, time series analysis, and many more. Thus, this benchmark represents a comprehensive suite of tests with which to evaluate automated machine learning systems.

[1] See https://gist.github.com/rhiever/27f795b00b95751ee38fd9e946c72b0b for a full list of building blocks.

[2] Benchmark data available at http://www.randalolson.com/data/benchmarks/.

14.4 Results

To provide an initial evaluation of TPOT-SI, we ran 30 replicates of TPOT-SI on the 160 supervised classification benchmark data sets described in Sect. 14.3.5. We compare these experiments to the same experiments with a version of TPOT without sensible initialization. In all cases, we measured pipeline accuracy as balanced accuracy [29], which corrects for class frequency imbalances in data sets by computing the accuracy on a per-class basis then averaging the per-class accuracies.

Figure 14.3 summarizes the difference in performance between TPOT-SI and TPOT on each benchmark. Overall, TPOT-SI showed no improvement on a large portion of the 160 benchmarks, small performance degradation on a few benchmarks, and fair improvement on a handful of benchmarks. Notably, the largest performance degradation was on the "tutorial" benchmark from [25] with a 5% median accuracy decrease, and the largest performance increase was on the "parity5" benchmark from [25] with a 12.5% median accuracy increase.

In order to provide better insight into why so many benchmarks saw no improvement with TPOT-SI, we plotted the original TPOT accuracy on each data

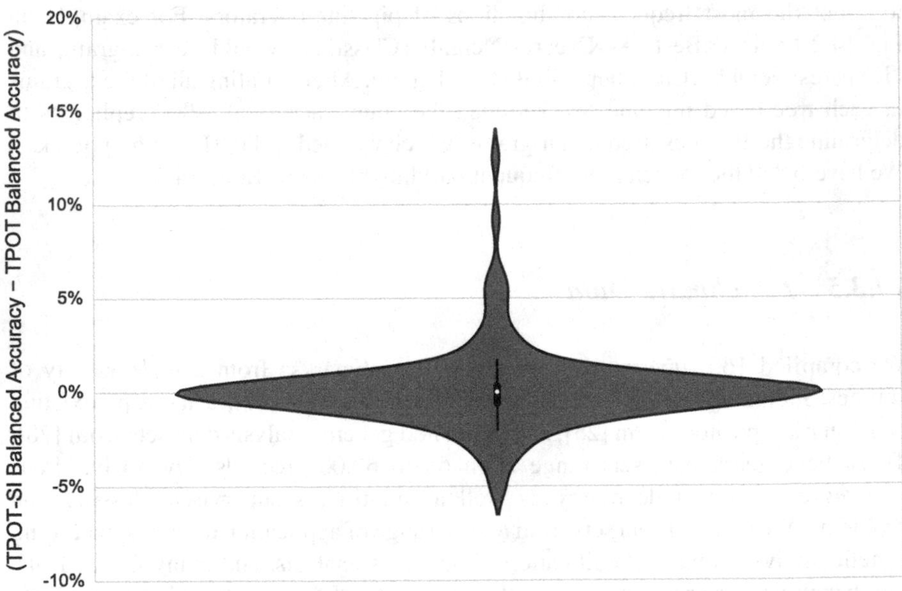

Fig. 14.3 Violin plot of the difference in median balanced accuracy between TPOT-SI and TPOT on the benchmarks. Positive values indicate an improvement in accuracy from TPOT-SI, whereas negative values indicate a degradation of accuracy from TPOT-SI. The width of the violin represents the relative density of points at that value, e.g., most differences are centered around 0% accuracy improvement. We note that the density is estimated from the underlying data, which is why it appears that there are differences in accuracy below −5%

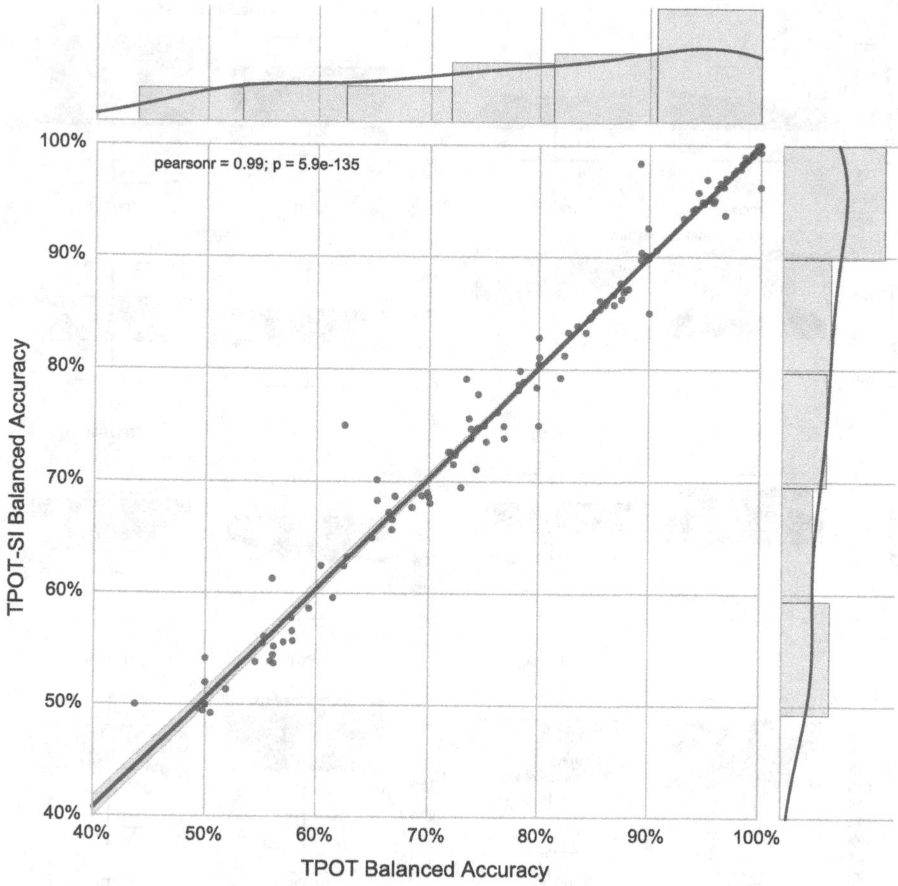

Fig. 14.4 Median balanced accuracy of TPOT-SI vs. the same for TPOT on the benchmarks. Each point represents the median balanced accuracies for one benchmark. The line represents a linear regression fit to the median accuracies, whereas the histograms on the sides show the density of the points on both axes

set vs. the TPOT-SI accuracy on the corresponding data set in Fig. 14.4. 49 of the benchmarks were already solved (>90% median balanced accuracy) with the base version of TPOT, which is why TPOT-SI saw no improvement on many of the benchmarks.

Finally, we show the distribution of balanced accuracies on the eight benchmarks with the largest performance differences in Fig. 14.5. Surprisingly, the only benchmark (out of all 160) with a statistically significant difference in performance is the "analcatdata_lawsuit" benchmark, where TPOT-SI allowed for a 9.2% higher accuracy on average. In the other benchmarks, TPOT-SI allowed for small but statistically insignificant improvements over TPOT.

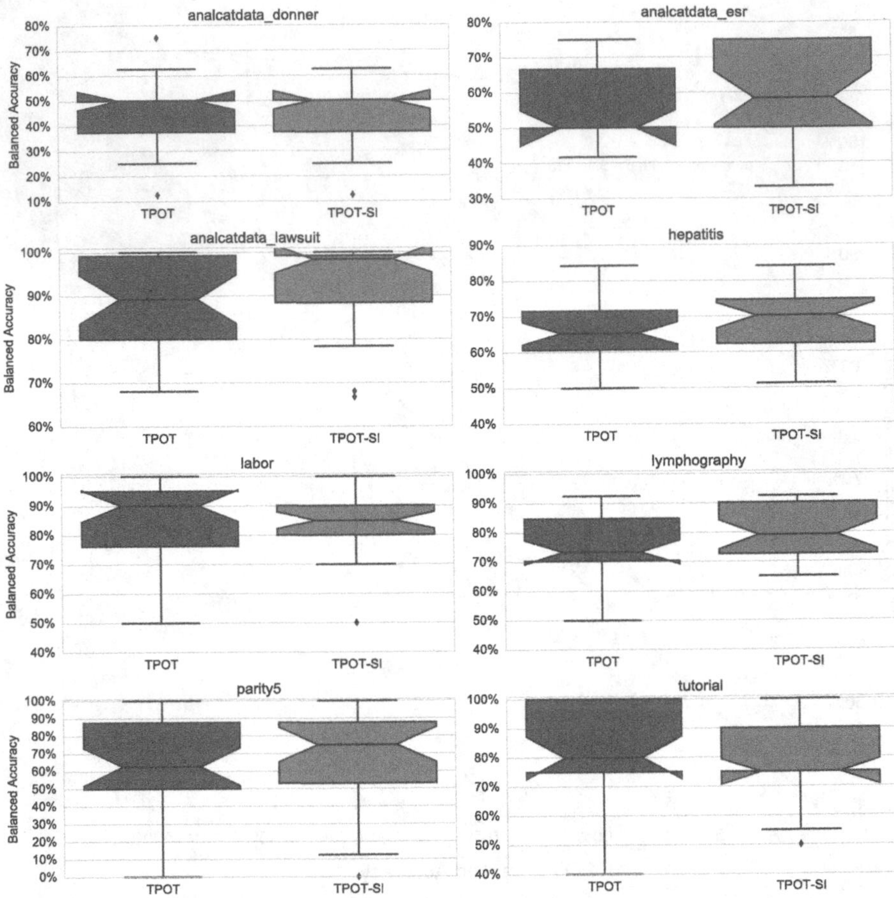

Fig. 14.5 Box plots showing the distribution of balanced accuracies for the eight benchmarks with the biggest difference in median accuracy between TPOT-SI and TPOT. Each box plot represents 30 replicates, the inner line shows the median, and the notches represent the bootstrapped 95% confidence interval of the median

14.5 Summary and Discussion

In this chapter, we presented preliminary results from implementing sensible initialization in TPOT. In summary, TPOT-SI saw significant improvement in performance on only one benchmark out of 160 (Fig. 14.5). Although our sensible initialization method could likely use improvement, we note that TPOT-SI did not significantly degrade performance on any of the benchmarks as well.

Of course, the key goal of this chapter extends beyond implementing a sensible initialization method in TPOT: We seek to identify the building blocks of machine learning pipelines, which is information that can be harnessed in many machine learning applications. In this chapter, we suggest that machine learning building blocks are small sequences of machine learning operators that occur frequently in pipelines used to solve benchmark classification problems. Although the building blocks that we identified do not seem to significantly improve performance via sensible initialization on many benchmarks, these results could be for many reasons. For example, that not all of the building blocks that we used are useful for all of the benchmarks, and in fact some could be detrimental on some benchmarks. We discuss how these limitations can be overcome in Sect. 14.6.

Furthermore, TPOT-SI automatically optimized pipelines for all 160 benchmarks, and discovered pipelines that achieve >90% median balanced accuracy on 52 of the benchmarks (and >80% on 79 of them) without any prior knowledge of the problem domains. These results show significant promise for GP-based automated machine learning systems. We note, however, that TPOT should not be considered a replacement for machine learning practitioners; rather, TPOT saves practitioners time by automating the tedious portions of machine learning pipeline design, but ultimately the practitioners will be responsible for deciding on the final pipeline. Similarly, we consider TPOT to be a "Data Science Assistant" and idea generator because it can discover unique ways to model data sets—and export its finding to the corresponding Python code—so practitioners can take TPOT pipelines and customize them for their particular application. To aid in the effort of providing an easily accessible GP-based automated machine learning system, we have released TPOT as an open source Python application at https://github.com/rhiever/tpot.

14.6 Looking Forward

The sensible initialization method implemented in TPOT-SI is quite simple, and there are many refinements that can be made. For example, we can use meta-learning techniques to intelligently match building blocks and pipeline configurations that will work well on the particular data set being analyzed [7]. In short, meta-learning uses information from previous machine learning runs to estimate how well each pipeline configuration will work on a particular data set. To place data sets on a standard scale, meta-learners compute meta-features from data sets, such as data set size, the number of features, and various aspects about the features, which are then used to map data set meta-features to corresponding pipeline configurations that work well on data sets with those meta-features. Such an intelligent meta-learning algorithm is likely to improve the TPOT sensible initialization process.

Similarly, we can harness expert knowledge about machine learning building blocks to bias the GP mutation and crossover operations, similar to what was done in [11]. In this case, we would provide the GP algorithm information about how well particular pipeline combinations perform on average, e.g., "Replacing a RandomForest operator with a DecisionTree operator is 89% likely to degrade accuracy." This information could then be used to bias the mutation and crossover operations toward producing better pipelines. Further, this information could be learned and refined over the optimization process, such that the GP algorithm will learn what makes an effective pipeline for the particular data set being analyzed.

Finally, population-based optimization methods such as GP are typically criticized for maintaining a large population of solutions, which can prove to be slow and wasteful for certain optimization problems. In this case, we can turn GP's purported weakness into a strength by creating an ensemble out of the GP populations. [3] explored this population ensemble method previously with standard GP and showed significant improvement, and it is a natural extension to create ensembles out of TPOT's population of machine learning pipelines.

In conclusion, automated machine learning is a field of research that is ripe for GP systems. We should focus our efforts on refining a GP-based automated machine learning system, and in particular highlight GP's strengths as compared to Bayesian optimization, simulated annealing, and greedy optimization techniques. TPOT represents our effort toward this goal, and we will continue to refine TPOT until it consistently produces human-competitive machine learning pipelines.

Acknowledgements We thank the Penn Medicine Academic Computing Services for the use of their computing resources. This work was supported by National Institutes of Health grants LM009012, LM010098, and EY022300.

References

1. Banzhaf, W., Nordin, P., Keller, R.E., Francone, F.D.: Genetic Programming: An Introduction. Morgan Kaufmann, San Meateo (1998)
2. Bergstra, J., Bengio, Y.: Random search for hyper-parameter optimization. J. Mach. Learn. Res. **13**, 281–305 (2012)
3. Bhowan, U., Johnston, M., Zhang, M., Yao, X.: Evolving diverse ensembles using genetic programming for classification with unbalanced data. Trans. Evol. Comput. **17**(3), 368–386 (2013)
4. Chen, T., Guestrin, C.: XGBoost: a scalable tree boosting system. CoRR abs/1603.02754 (2016). http://arxiv.org/abs/1603.02754
5. Deb, K., Pratap, A., Agarwal, S., Meyarivan, T.: A fast and elitist multiobjective genetic algorithm: NSGA-II. IEEE Trans. Evol. Comput. **6**, 182–197 (2002)
6. Feurer, M., Klein, A., Eggensperger, K., Springenberg, J., Blum, M., Hutter, F.: Efficient and robust automated machine learning. In: Cortes, C., Lawrence, N., Lee, D., Sugiyama, M., Garnett, R. (eds.) Advances in Neural Information Processing Systems 28, pp. 2944–2952. Curran Associates, Inc., Red Hook (2015)
7. Feurer, M., Springenberg, J.T., Hutter, F.: Initializing bayesian hyperparameter optimization via meta-learning. In: Proceedings of the 29th AAAI Conference on Artificial Intelligence, January 25–30, 2015, Austin, pp. 1128–1135 (2015)
8. Fortin, F.A., De Rainville, F.M., Gardner, M.A., Parizeau, M., Gagné, C.: DEAP: evolutionary algorithms made easy. J. Mach. Learn. Res. **13**, 2171–2175 (2012)
9. Garca-Arnau, M., Manrique, D., Ros, J., Rodrguez-Patn, A.: Initialization method for grammar-guided genetic programming. Knowl.-Based Syst. **20**, 127–133 (2007). The 26th SGAI International Conference on Innovative Techniques and Applications of Artificial Intelligence
10. Goldberg, D.E.: The Design of Innovation: Lessons from and for Competent Genetic Algorithms. Kluwer Academic Publishers, Norwell (2002)
11. Greene, C.S., White, B.C., Moore, J.H.: An expert knowledge-guided mutation operator for genome-wide genetic analysis using genetic programming. In: Pattern Recognition in Bioinformatics, pp. 30–40. Springer, Berlin (2007)

12. Greene, C.S., White, B.C., Moore, J.H.: Sensible initialization using expert knowledge for genome-wide analysis of epistasis using genetic programming. In: 2009 IEEE Congress on Evolutionary Computation, pp. 1289–1296 (2009)
13. Hastie, T.J., Tibshirani, R.J., Friedman, J.H.: The Elements of Statistical Learning: Data Mining, Inference, and Prediction. Springer, New York (2009)
14. Hutter, F., Lücke, J., Schmidt-Thieme, L.: Beyond manual tuning of hyperparameters. Künstl. Intell. **29**, 329–337 (2015)
15. Kanter, J.M., Veeramachaneni, K.: Deep feature synthesis: towards automating data science endeavors. In: Proceedings of the International Conference on Data Science and Advance Analytics. IEEE, Piscataway (2015)
16. Koza, J.R.: Genetic Programming: On the Programming of Computers by Means of Natural Selection. MIT Press, Cambridge (1992)
17. Lichman, M.: UCI machine learning repository (2013). http://archive.ics.uci.edu/ml
18. Luke, S., Panait, L.: A survey and comparison of tree generation algorithms. In: Spector, L., Goodman, E.D., Wu, A., Langdon, W.B., Voigt, H.M., Gen, M., Sen, S., Dorigo, M., Pezeshk, S., Garzon, M.H., Burke, E. (eds.) Proceedings of the 6th Genetic and Evolutionary Computation Conference, GECCO '01, pp. 81–88. Morgan Kaufmann, San Francisco (2001)
19. Martinsson, P.G., Rokhlin, V., Tygert, M.: A randomized algorithm for the decomposition of matrices. Appl. Comput. Harmon. Anal. **30**, 47–68 (2011)
20. Olson, R.S., Bartley, N., Urbanowicz, R.J., Moore, J.H.: Evaluation of a tree-based pipeline optimization tool for automating data science (2016). Arxiv e-print. http://arxiv.org/abs/1603.06212
21. Olson, R.S., Urbanowicz, R.J., Andrews, P.C., Lavender, N.A., Kidd, L.C., Moore, J.H.: Automating biomedical data science through tree-based pipeline optimization. In: Applications of Evolutionary Computation: 19th European Conference, EvoApplications 2016, Porto, March 30 April 1, 2016, Proceedings, Part I, pp. 123–137. Springer International Publishing, Cham (2016)
22. O'Neill, M., Ryan, C.: Grammatical Evolution: Evolutionary Automatic Programming in a Arbitrary Language. Genetic Programming, vol. 4. Kluwer Academic Publishers, Dordrecht (2003)
23. Pedregosa, F., Varoquaux, G., Gramfort, A., Michel, V., Thirion, B., Grisel, O., Blondel, M., Prettenhofer, P., Weiss, R., Dubourg, V., Vanderplas, J., Passos, A., Cournapeau, D., Brucher, M., Perrot, M., Duchesnay, E.: Scikit-learn: machine learning in Python. J. Mach. Learn. Res. **12**, 2825–2830 (2011)
24. Poli, R., Langdon, W.B., McPhee, N.F.: A Field Guide to Genetic Programming. Lulu Enterprises, UK Ltd, Egham (2008)
25. Reif, M.: A comprehensive dataset for evaluating approaches of various meta-learning tasks. In: First International Conference on Pattern Recognition and Methods (ICPRAM) (2012)
26. Simon, P.: Too Big to Ignore: The Business Case for Big Data. Wiley & SAS Business Series. Wiley, New Delhi (2013)
27. Snoek, J., Larochelle, H., Adams, R.P.: Practical bayesian optimization of machine learning algorithms. In: Pereira, F., Burges, C.J.C., Bottou, L., Weinberger, K.Q. (eds.) Advances in Neural Information Processing Systems 25, pp. 2951–2959. Curran Associates, Inc., Red Hook (2012)
28. Urbanowicz, R.J., Kiralis, J., Sinnott-Armstrong, N.A., Heberling, T., Fisher, J.M., Moore, J.H.: GAMETES: a fast, direct algorithm for generating pure, strict, epistatic models with random architectures. BioData Min. **5**, 16 (2012)
29. Velez, D.R., et al.: A balanced accuracy function for epistasis modeling in imbalanced datasets using multifactor dimensionality reduction. Genet. Epidemiol. **31**(4), 306–315 (2007)
30. Zutty, J., Long, D., Adams, H., Bennett, G., Baxter, C.: Multiple objective vector-based genetic programming using human-derived primitives. In: Proceedings of the 2015 Annual Conference on Genetic and Evolutionary Computation, GECCO '15, pp. 1127–1134. ACM, New York (2015)

Index

R. Riolo et al. (eds.), *Genetic Programming Theory and Practice XIV*, Genetic and Evolutionary Computation,
https://doi.org/10.1007/978-3-319-97088-2

Printed in the United States
By Bookmasters